Petroleum and the Environment

Edith Allison and Ben Mandler

Petroleum and the Environment
Edith Allison and Ben Mandler
ISBN: 978-1721175468

© 2018 American Geosciences Institute.

Published and printed in the United States of America. All rights reserved. No part of this work may be reproduced or transmitted in any form or by any means, electronic or mechanical, recording, or any information storage and retrieval system without the expressed written consent of the publisher.

Design by Brenna Tobler, AGI Graphic Designer, and Ben Mandler, AGI Critical Issues Program.

For more information on the American Geosciences Institute and its publications, check us out at store.americangeosciences.org.

Contact:
Ben Mandler, Schlumberger Senior Researcher
American Geosciences Institute
4220 King Street, Alexandria, VA 22302
www.americangeosciences.org
bmandler@americangeosciences.org
(703) 379-2480, ext. 226

Cover photos: clockwise from top-left: Oil Rig, Williston Oil Field, North Dakota, Lindsey Gira, CC BY 2.0 via Wikimedia Commons; Alyeska Pipeline at mile 562, Alaska, Copyright © Larry Fellows, Arizona Geological Survey; Los Angeles highways, ©Shutterstock.com/S. Borisov; Petrochemical Plant, Corpus Christi, Texas, ©Shutterstock.com/Trong Nguyen. Cover background: ©Shutterstock.com/Sergey Nivens.

AGI Critical Issues Program: www.americangeosciences.org/critical-issues
Supported by the AAPG Foundation. © 2018 American Geosciences Institute

Acknowledgments

This publication benefited from the expertise of many reviewers from different sectors all over the United States. Thank you to our internal AGI reviewers: Allyson Anderson Book, Maeve Boland, Kelly Kryc, and Cassaundra Rose; and to our external reviewers: Donna S. Anderson, Kevin Book, James Cooper, Carol Frost, Ryan Hill, David Houseknecht, Kerry Klein, Jennifer Miskimins, Jacques Rousseau, Steven Schamel, and four anonymous reviewers. Additional information from Timothy Zebulske (Bureau of Land Management) helped in the production of the Pinedale case study, and Tracey Mercier (U.S. Geological Survey) provided higher-resolution graphics for Part 5: Using Produced Water.

Thank you to the California Department of Public Health, the Environmental Defense Fund, Northeast Natural Energy, the Oklahoma Energy Resources Board, the Union of Concerned Scientists, the Utah Geological Survey, and the Texas Oil and Gas Association for permission to reproduce additional images.

And finally, thanks to Bridget Scanlon, David Houseknecht, John Fontana, Katherine Saad, Sherilyn Williams-Stroud, Steven Dade, and Susan Nissen for kindly sharing their stories in Part 24: Geoscientists in Petroleum and the Environment.

This work was supported by the American Association of Petroleum Geologists Foundation.

Foreword

The American Geosciences Institute (AGI) represents and serves the geoscience community by providing collaborative leadership and information to connect Earth, science, and people. In 2003, AGI, in cooperation with our member societies and several federal agencies, released a publication called *Petroleum and the Environment* to give the general public, educators, and policy makers a better understanding of environmental concerns related to petroleum resources.

In the years that followed, the shale boom yielded a huge increase in domestic energy production, which came with new environmental challenges, opportunities, and uncertainties. We realized that our 2003 publication no longer provided the breadth or depth of information needed to understand how today's oil and gas operations interact with the environment.

This update to the original *Petroleum and the Environment* provides a completely rewritten overview of many environmental considerations associated with current practices as of 2018 in oil and gas exploration, production, processing, refining, transportation, and use. Designed as a series of factsheets that can be read individually or as a single publication, online or in print, this update is an accessible and impartial explanation of a complex subject. It can serve as a starting point for someone interested in learning more about petroleum and the environment, or as a reference work.

We gratefully acknowledge funding from the American Association of Petroleum Geologists Foundation, which made this publication possible. I would also like to recognize Edie Allison, an experienced petroleum geologist and writer, and Ben Mandler, from AGI, who collaborated to write and produce this revised version of *Petroleum and the Environment*.

Eve Sprunt
President, American Geosciences Institute, 2017-2018

Foreword

For more than fifty years the American Association of Petroleum Geologists (AAPG) Foundation has focused on advancing the science of geology, particularly petroleum geology, by supporting educational and research programs to benefit the geoscience profession and the public.

Petroleum, both oil and natural gas, form the base of our energy infrastructure, which provides the foundation for modern society. There is a strong correlation between energy consumption and economic growth. And as nations seek to provide better lives for their citizens and coming generations, it's essential that they have access to affordable, reliable energy supplies developed and produced in an environmentally-responsible manner.

The AAPG Foundation is pleased to support this publication by the American Gesociences Institute that seeks to enhance public understanding of petroleum resources and their relation to environmental, human, and societal health.

David Curtiss
Executive Director, American Association of Petroleum Geologists Foundation

Contents

Part	Title
1	Petroleum and the Environment: an Introduction
2	Water in the Oil and Gas Industry
3	Induced Seismicity from Oil and Gas Operations
4	Water Sources for Hydraulic Fracturing
5	Using Produced Water
6	Groundwater Protection In Oil and Gas Production
7	Abandoned Wells
8	What Determines the Location of a Well?
9	Land Use in the Oil and Gas Industry
10	The Pinedale Gas Field, Wyoming
11	Heavy Oil
12	Oil and Gas in the U.S. Arctic
13	Offshore Oil and Gas
14	Spills in Oil and Natural Gas Fields
15	Transportation of Oil, Gas, and Refined Products
16	Oil Refining and Gas Processing
17	Non-Fuel Products of Oil and Gas
18	Air Quality Impacts of Oil and Gas
19	Methane Emissions in the Oil and Gas Industry
20	Mitigating and Regulating Methane Emissions
21	U.S. Regulation of Oil and Gas Operations
22	Health and Safety in Oil and Gas Extraction
23	Subsurface Data in the Oil and Gas Industry
24	Geoscientists in Petroleum and Environment
	Glossary of Terms
	References

Petroleum and the Environment
Part 1

Petroleum and the Environment: an Introduction
Relationships between oil and gas and the environment in historical context

Introduction

When oil and gas were first extracted and used on an industrial scale in the 19th century, they provided significant advantages over existing fuels: they were cleaner, easier to transport, and more versatile than coal and biomass (wood, waste, and whale oil). Diesel and gasoline derived from oil revolutionized the transportation sector. Through developments in chemical engineering, oil and gas also provided the raw materials for a vast range of useful products, from plastics to fertilizers and medicines. By the 20th century, oil and gas had become essential resources for modern life: as both fuel and raw material, the versatility and abundance of oil and gas helped to facilitate unprecedented economic growth and improved human health around the world.

Despite rapid advances in renewable energy technologies, in 2016 oil and gas accounted for two thirds of U.S. energy consumption[1] and over half of all the energy consumed worldwide.[2] Annual U.S. oil and gas production is expected to increase beyond 2040.[3] Developments in policy, technology, and public opinion may change this trend, but oil and gas will likely play a fundamental role in U.S. and global energy production and consumption for much of the 21st century.

How can the environmental and public health risks of the energy sector be minimized while ensuring a consistent and abundant energy supply? An important step in addressing environmental and health concerns is understanding the risks and the options

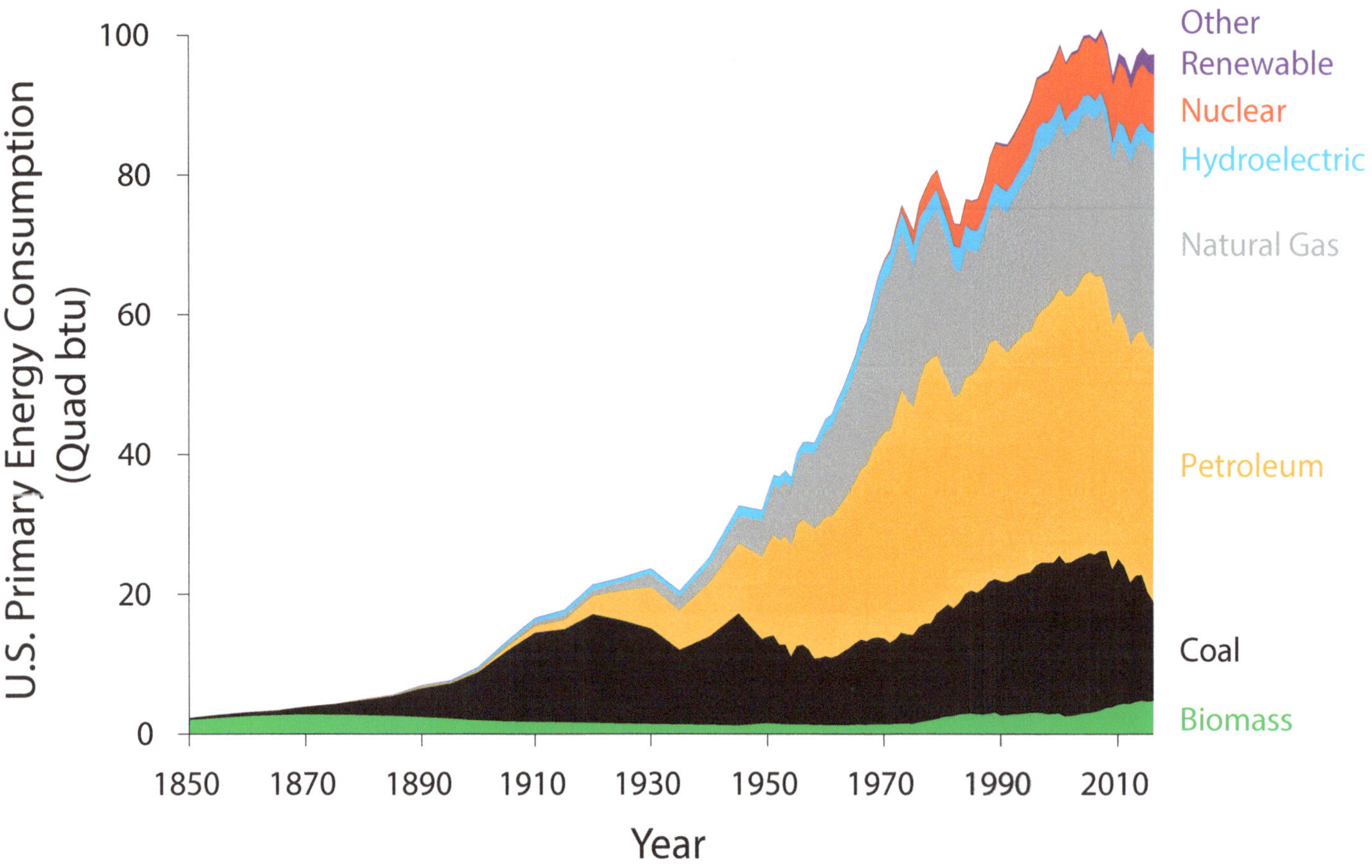

Sources of energy used in the United States, 1850-2016. "Other Renewable" includes solar, wind, and geothermal energy. Image credit: American Geosciences Institute. Data source: Energy Information Administration.[4,5]

Petroleum and the Environment
Part 1: Petroleum and the Environment: an Introduction

available to reduce them. In this series, 24 factsheets and case studies provide an overview of the many intersections between the oil and gas industry and the environment. Specific attention is paid to: (1) environmental and human health issues in the exploration, production, refining and processing, and transportation of oil and natural gas; and (2) some of the technologies, management practices, and regulations that can help to address these issues.

A Note on Climate Change

The combustion of fossil fuels (coal, oil, and natural gas) releases large quantities of carbon dioxide (and other greenhouse gases) into the atmosphere, which has a wide range of environmental impacts. The full extent of these impacts is not yet known, but they include rising global temperatures, ocean acidification, sea level rise, and a variety of other impacts on weather, natural hazards, agriculture, and more, many of which are likely to increase into the future.[10,11,12] While agriculture and land use change also emit carbon dioxide and other greenhouse gases (especially methane), fossil fuels, especially coal and oil, produce the majority of anthropogenic (human-caused) emissions of greenhouse gases on a global scale.[13]

Since peaking in 2007, U.S. emissions of greenhouse gases have decreased largely due to changes in electric power generation (decreased electricity demand and decreased use of coal for electricity generation in favor of natural gas and, more recently, renewables).[14] With continued changes in how energy is generated and used, emissions from the energy sector as a whole may continue to decrease. Meanwhile, there are actions that can be taken to reduce the carbon emissions from the oil and gas sector, such as reducing gas leaks, using less energy-intensive exploration and production techniques, and capturing and storing carbon emissions. These efforts are addressed throughout this series where relevant to each subtopic. For those interested in learning more about climate change and its relationship to the combustion of fossil fuels as an energy source, a selection of resources is provided at the end of this introduction.

Recent Developments

Oil and gas exploration, production, and use have radically changed since the beginning of the 21st century. The use of horizontal drilling with hydraulic fracturing to access previously uneconomic oil and gas deposits led to unprecedented increases in oil and gas production: from 2006 to 2015, U.S. natural gas production increased by 40%,[6] while from 2008 to 2015, U.S. oil production increased by 88%.[7] This growth in production has led to commensurate growth in oil and gas transportation, processing and refining, use in agriculture and manufacturing, and energy exports.[8,9]

However, the recent growth in oil and gas production has increased or renewed some longstanding concerns over their impact on the environment, while also giving rise to some new concerns. Areas of major change and/or public concern include:

- **Hydraulic fracturing ("fracking")** – this technique of fracturing rocks to extract oil and gas has been used since the 1940s, but its combination with horizontal drilling to extract oil and gas from shale led to a surge in hydraulic fracturing starting around 2005. The widespread use of hydraulic fracturing has raised questions about the large amount of water used in the process, which may compete with other fresh water demands in some areas, and has motivated research into alternative fluids. Hydraulic fracturing has also highlighted the issue of groundwater

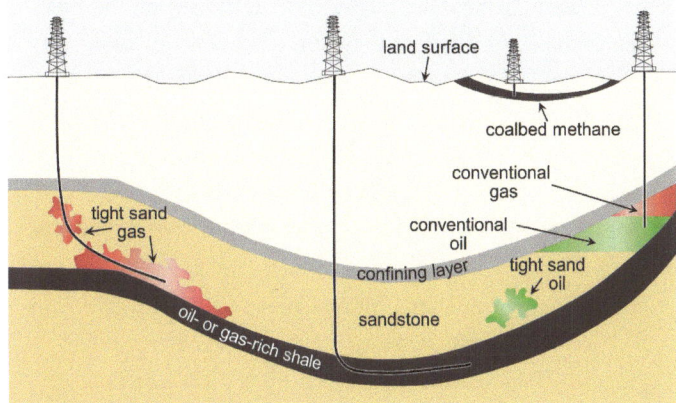

Schematic showing the various types of oil and gas deposits. Recent advances in directional (especially horizontal) drilling and hydraulic fracturing have led to substantial increases in production from shale as well as tight oil and gas sandstone. Image credit: U.S. Environmental Protection Agency.[15]

Petroleum and the Environment
Part 1: Petroleum and the Environment: an Introduction

protection, partly due to concerns over the fracturing process itself and partly due to the use of toxic chemicals in some hydraulic fracturing fluids. This adds a new element of concern to a longstanding problem: old or poorly constructed wells may leak a variety of fluids if the cement or steel portions of the well are compromised, whether they are hydraulically fractured or not. Identifying instances and sources of groundwater contamination is an ongoing challenge for research scientists, regulators, and industry.

- **Induced earthquakes** – many human activities can trigger earthquakes, including geothermal energy production, filling up reservoirs behind dams, and groundwater extraction.[16,17] Oil and gas operations can trigger earthquakes through two main processes: underground wastewater injection and hydraulic fracturing. The largest induced earthquakes from oil and gas operations have been caused by the underground injection of large volumes of wastewater extracted along with oil and gas ("produced water").[18] This water is often too salty to release into surface waterbodies so it is instead injected deep underground, where it can increase the likelihood of earthquakes on existing faults. Hydraulic fracturing very rarely causes noticeable earthquakes, but it has triggered some small but noticeable earthquakes in parts of the United States and Canada.[19] Some states, particularly Oklahoma and Kansas, have observed a decrease in induced seismicity since 2015 due to decreased oil production (and therefore less wastewater in need of disposal) and new regulations constraining wastewater injection volumes and rates.[20]

- **Land use** – advances in horizontal drilling mean that wells don't need to be placed directly above a resource, so the location of well sites can be planned to reduce their surface impact, and multiple wells can be drilled in different directions from a single site. However, the boom in horizontal drilling and hydraulic fracturing has led to increased oil and gas activity in many areas, including some areas that had previously had little activity, resulting in increased overall land disturbance in some parts of the country.

- **Methane emissions** – the surge in U.S. natural gas production has led to natural gas replacing coal as the largest source of electricity in the United States.[21] Burning natural gas releases less carbon dioxide than coal, so in this sense the transition from coal to natural gas has had a positive environmental impact. However, methane, the main component of natural gas, is itself a potent greenhouse gas, so any leaks during natural gas production and distribution will partially offset this benefit. Improved monitoring of methane emissions, targeted repair and replacement of equipment, and potential regulations all play a role in minimizing methane leaks.

- **Heavy oil and oil sands** – some of the largest oil resources in the world consist of thick, heavy oil that is extracted and processed with specific, energy-intensive techniques. In the United States, California has long produced heavy oil from the Kern River Oil Field outside Bakersfield,[22] but the largest heavy oil producers are Venezuela and Canada. In Canada, a significant proportion of oil production comes from oil sands (also known as "tar sands"), which are a mixture of clay, sand, water, and bitumen (a very thick oil). Canadian oil sand production has increased substantially since 2005. Deeper oil sand deposits are commonly extracted by heating the oil, which thins it so that it can flow up through a well. Shallower deposits are extracted by open-pit mining of the oil sands, which are then processed to remove the oil. Regardless of the production technique, oil sand production is energy-intensive and therefore results in higher overall emissions of carbon dioxide and combustion-related air pollutants. Open-pit mining of oil sands in particular presents additional environmental concerns, such as risks to air and water quality from dust and waste ponds.

- **Transportation and spills** – increases in oil and gas production and consumption require enhanced transportation infrastructure. About 90% of crude oil and refined products, and essentially all natural gas, are transported through millions of miles of (mostly underground) pipelines. Spills of crude oil and refined products represent less than 0.001% of the total amount transported, but this small percentage amounts to millions of gallons spilled each year.[23] Most spills are small but some can have significant local impacts and require extensive and expensive cleanup efforts.

Petroleum and the Environment
Part 1: Petroleum and the Environment: an Introduction

- **Offshore drilling** – advances in offshore drilling technology have allowed oil and gas to be produced in increasingly deep water. These extreme conditions pose particular technological and environmental challenges. For example, in 2010, a defective well in the Macondo prospect of the Gulf of Mexico caused the largest marine oil spill in history and killed 11 workers on the Deepwater Horizon drilling rig. Since this spill, regulations and industry practices have changed substantially to reduce the environmental risks of offshore oil and gas production, but many concerns remain.

The Purpose of This Series

This series of factsheets and case studies discusses developments in the topics listed above, as well as some less visible topics, such as the importance of data, non-fuel products of oil and gas, the positive and negative impacts of refining and processing, and the many factors that determine the location of a well. Our aim is to provide an overview of the many ways in which the oil and gas industry interacts with people and the environment, including major risks and hazards; progress that has been made in addressing these issues; and how the geosciences, technology, and regulations are used in attempts to balance the need for abundant, affordable energy with the need to protect and preserve environmental and human health.

The information provided herein represents the most current reliable information available to the authors at the time of publishing (Spring 2018). Data relating to petroleum and the environment change through time: government statistics may be updated weekly, monthly, annually, or less frequently; research on aspects of energy and the environment is regularly published in peer-reviewed journals; and regulations affecting energy exploration, production and use are developed and/or legally contested in an ongoing but slow process. Details that are more likely to change in the near future or are associated with substantial uncertainty are indicated. Every attempt has been made to provide references and recommend additional resources that are freely available online, to allow the interested reader to dive deeper into each topic. Not all topics have been covered in the same level of detail: for some topics for which high-quality, accessible resources are freely available (e.g., oil spills in marine settings), the reader is directed toward these existing resources for further information.

References & More Resources

For a complete listing of references, see the "References" section of the full publication, *Petroleum and the Environment*, or visit the online version at: www.americangeosciences.org/critical-issues/petroleum-environment

Petroleum and the Environment (2018) is a completely rewritten update of AGI's 2003 publication, *Petroleum and the Environment*. You can access the 2003 publication at https://www.agiweb.org/environment/publications/petroleum.pdf

Additional resources are provided in each of the 24 sections of this series where they are most topically relevant. Below is a selection of resources that relate to petroleum and the environment but are not provided elsewhere in this series.

Intergovernmental Panel on Climate Change – Fifth Assessment Report. http://www.ipcc.ch/report/ar5/

U.S. Environmental Protection Agency – Global Greenhouse Gas Emissions Data. https://www.epa.gov/ghgemissions/global-greenhouse-gas-emissions-data

International Energy Agency - World Energy Flows Sankey Diagram. https://www.iea.org/Sankey/

Institute of Physics (2008). The Role of Physics in Improving the Efficiency of Electricity Generation and Supply. https://www.iop.org/publications/iop/2008/file_38218.pdf

World Resources Institute – Resource Watch. https://resourcewatch.org/

International Energy Agency (2017). CO_2 Emissions from Fuel Combustion 2017. http://www.iea.org/bookshop/757-CO2_Emissions_from_Fuel_Combustion_2017

Raimi, D. (2017). The Fracking Debate: The Risks, Benefits, and Uncertainties of the Shale Revolution. Columbia University Press, 280 pp.

Petroleum and the Environment
Part 2

Water in the Oil and Gas Industry
An overview of the many roles of water in oil and gas operations

Introduction
The oil and gas industry consumes and produces water. Water is used to drill and hydraulically fracture ("frack") wells, refine and process oil and gas, and produce electricity in some natural gas power plants. Water is also naturally present in the rocks that contain oil and gas and is extracted along with the oil and gas as "produced water", sometimes in large quantities. The quantity and quality of water used, produced, and disposed of or re-used varies enormously depending on local geology, financial constraints, and regulations, with implications for the environmental impacts of oil and gas production.

Many aspects of water use in the oil and gas industry are covered in more detail elsewhere in this series – see the list at the end of this section for more information on each.

Sourcing Water
Water used in the production of oil and gas is often locally sourced from groundwater, rivers, or lakes (both natural and artificial). Where fresh water is in high demand for other uses, water reuse and alternative water sources (e.g., brackish groundwater) are attractive options.

Transporting Water
Water is often transported by trucks, which bring water to the oil or gas well for drilling and hydraulic fracturing, and take used or produced water away for treatment, reuse, and disposal. In areas with many established wells, pipelines may be installed to transport water, improving efficiency and safety, and decreasing traffic.

Using Water
Water is used during drilling to lubricate and cool the drill and remove drilling mud and rock debris. For hydraulic fracturing operations, water is mixed with chemicals that improve its ability to create fractures in the rock, and with sand to hold the fractures open and allow oil or gas to flow into the well. Although most wells do not leak, some old or poorly constructed wells may pose a contamination risk to nearby groundwater supplies.

Quick Facts: Water Volumes

In 2010, total U.S. water use was about 355 billion gallons per day, or 1,100 gallons per person per day.[1] Major water users were:
- Thermoelectric power: 45%
- Irrigation: 38%
- Public supply: 14%
- Mining, including oil and gas extraction: 2%

Hydraulic fracturing water use per well varies from about 1.5 million gallons to about 16 million gallons.[2] For comparison, a million gallons would cover a football field to a depth of almost 4 feet.

In the U.S., 2.5 billion gallons of produced water are extracted along with oil and gas every day,[3] most of which is then injected underground, either to enhance oil recovery or simply for disposal.[4]

Tanks like these in the Fayetteville Shale area (Arkansas) are commonly used to supply the water required for hydraulic fracturing operations. Image credit: Bill Cunningham, USGS.

Petroleum and the Environment
Part 2: Water in the Oil and Gas Industry

Produced Water

Water produced along with oil and gas is often naturally salty and may contain oil residues, chemicals from hydraulic fracturing and drilling fluids, and natural contaminants from the rocks themselves. It is usually either disposed of deep underground or treated and reused, though some is allowed to partially evaporate in surface pits. The amount of water produced by a well can vary from almost none to over 100 barrels of water per barrel of oil. Nationally, an average of about 10 barrels of water are produced for each barrel of oil.[3]

Water Treatment, Reuse, and Disposal

Produced water must be either re-used or disposed of. Re-use typically requires some treatment to remove oil residues, salts, and other chemicals, depending on how the water will be reused. In some cases, produced water is temporarily stored in surface pits to evaporate some of the water. This can affect local air quality, and if pits leak, they can contaminate groundwater supplies. In many places, large amounts of produced water are disposed of through deep underground injection wells. This has triggered earthquakes in Oklahoma, Kansas, and some other parts of the country. Access to disposal wells, earthquake prevention, water needs for other wells, produced water volumes, and treatment costs are all important factors when deciding how to dispose of or treat and reuse produced water.

Changes in Water Use

Although some hydraulic fracturing ("fracking") has been used since the 1940s, the boom in hydraulic fracturing since 2005 (especially using multiple frack treatments within single wells) has changed how water is used in oil and gas production. Fracked and non-fracked oil wells use similar amounts of water over their lifetime,[6] but the timing of water use is different. Water use increases over the life of a non-fracked well – large volumes of water may be injected into older wells to push out additional oil, a process called water flooding. Conversely, water use in many hydraulically fractured wells is very high at first but often decreases over time. Many hydraulically fractured gas wells use more water than non-fracked gas wells,[7] although water use varies substantially between different wells in different places.[6]

Regulation of Water

In general, individual states regulate oil and gas operations on state and private land, while federal agencies oversee operations on federal lands. Underground injection of produced water is regulated by the U.S. Environmental Protection Agency (EPA) or by states to which the EPA has delegated authority. The regulation and ownership of water in the U.S. varies greatly from place to place, but all water withdrawals from public or private water sources require approval from the relevant owner and/or regulatory agency.

Wastewater is often transported by truck to disposal facilities such as this one near Platteville, Colorado. After some treatment, most wastewater is disposed of deep underground. Image credit: William Ellsworth, USGS.[5]

References & More Resources

For a complete listing of references, see the "References" section of the full publication, *Petroleum and the Environment*, or visit the online version at: www.americangeosciences.org/critical-issues/petroleum-environment

Water in the Oil and Gas Industry: elsewhere in this series:

Part 3: Induced Seismicity in the Oil and Gas Industry
Part 4: Water Sources for Hydraulic Fracturing
Part 5: Using Produced Water
Part 6: Groundwater Protection in Oil and Gas Production
Part 14: Spills in Oil and Natural Gas Fields
Part 15: Transportation of Oil, Gas, and Refined Products
Part 21: U.S. Regulation of Oil and Gas Operations

Petroleum and the Environment
Part 3

Induced Seismicity from Oil and Gas Operations
Earthquakes caused by wastewater disposal and hydraulic fracturing

Manmade Earthquakes

Any activity that significantly changes the pressure on or fluid content of rocks has the potential to trigger earthquakes. This includes geothermal energy production, water storage in large reservoirs, groundwater extraction, underground injection of water for enhanced oil recovery, and large-scale underground disposal of waste liquids.[1] The fact that underground fluid injection can trigger damaging earthquakes has been understood since the 1960s, but historically such earthquakes have been very rare.[2] The sharp rise in noticeable earthquakes in the central United States from 2008 to 2015 was caused by massive increases in the underground disposal of produced water from the oil and gas industry.[3] Since mid-2015, declining rates of produced water disposal have led to fewer earthquakes in the central United States.

Hydraulic fracturing has caused some small earthquakes, but these are comparatively unusual: in a 2016 study of Canadian wells, 0.3% of hydraulically fractured wells were linked to earthquakes of at least magnitude 3.[4]

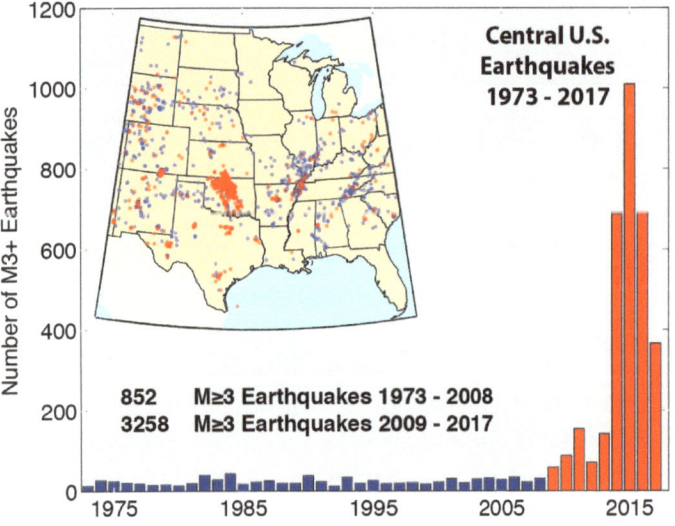

The sharp rise in small but noticeable earthquakes in the central United States, caused largely by huge increases in the underground disposal of produced water from oil and gas production. M = magnitude (see blue box). Image credit: U.S. Geological Survey[5]

Earthquakes Caused by Wastewater Disposal

In the central United States, particularly in central Oklahoma and south-central Kansas, the oil and gas boom in the early 2000s vastly increased the amount of produced water in need of disposal. Across the region, as more water was injected in deep disposal wells, earthquake activity increased:

- In Oklahoma, wastewater disposal rates tripled from 1 million barrels per day in 2010 to almost 3 million in 2014. Before 2008, Oklahoma had only a few earthquakes larger than magnitude 3 (M>3) per year; in 2014 the state had 579; and in 2015 there were 903 M>3 earthquakes in Oklahoma. These earthquakes clustered in areas with many large-volume disposal wells, strengthening the link between underground water disposal and induced earthquakes.[3] In 2015, underground water disposal began to decline, and in 2016 the number of M>3 earthquakes decreased to 623. This decline in underground disposal of produced water reflects both reduced production (due to lower oil prices) and state regulations.[3]

Induced Seismicity: Fast Facts and Figures

- 2.5 billion gallons of water are produced every day in the U.S. from about 900,000 oil and gas wells.[6] That's 7.5 gallons per U.S. resident per day.
- Roughly 90% of all produced water is injected[6] into roughly 150,000 wells: 40,000 are used only for wastewater disposal and the rest inject water for enhanced oil recovery.[7]
- Earthquakes caused by underground wastewater disposal have been most common and powerful in Oklahoma.[8]
- In Oklahoma, earthquake activity peaked in early 2015, with roughly three M>3 earthquakes every day.[9] From early 2015 to early 2017, earthquake activity decreased to less than one M>3 earthquake per day as less wastewater was injected underground.[3]

Petroleum and the Environment
Part 3: Induced Seismicity from Oil and Gas Operations

Damage caused by the magnitude 5.7 earthquake near Prague, Oklahoma, November 6, 2011. Unreinforced stone and brick buildings (especially chimneys) are some of the most vulnerable structures in any earthquake. Image credit: Brian Sherrod, USGS[10]

- In south-central Kansas, deep wastewater disposal volumes increased roughly tenfold from 2011 to 2014. In Harper and Sumner counties, which historically do not experience M>3 earthquakes in most years, over 100 M>3 earthquakes were recorded in 2015.[11] From 2015, decreasing injection volumes were followed by a decrease in earthquakes, and in 2016 there were fewer than 20 M>3 earthquakes in all of Kansas.[12] In 2017, Kansas saw slightly more earthquakes, but overall earthquake activity remained low.
- Texas went from having roughly two M>3 earthquakes per year before 2008 to around 12 per year from 2011 to 2016.[13]

Other areas that have seen noticeable earthquakes induced by wastewater disposal (in much lower numbers than Oklahoma) include Colorado, New Mexico, Arkansas, Ohio, and Wyoming.

Earthquakes Caused by Hydraulic Fracturing

Hydraulic fracturing does not generally cause earthquakes large enough to be felt (M>3), but there have been some exceptions:

- A study in Canada linked roughly 0.3% of hydraulically fractured wells to M>3 earthquakes.[4] Although most of these earthquakes occur close to and at the same time as hydraulic fracturing operations, a small percentage of induced earthquakes may occur months later.[14]
- In Ohio, both Poland Township (2014) and Harrison County (2015) have experienced M3 earthquakes caused by hydraulic fracturing.[3,15]
- In Oklahoma, some small (mostly ≤ M3) earthquakes have been linked to hydraulic fracturing in a small proportion of hydraulically fractured wells.[3]

Earthquake Risk Management and Mitigation by States

State regulators focus on identifying the precise location and magnitude of an earthquake and then determining its cause. If earthquakes are linked to wastewater injection, regulators can instruct operators to cease or limit injection rates and water volumes in nearby wells.[16,17] Many regulators also require that new injection wells avoid areas near known active faults. In Oklahoma, these techniques have effectively reduced the number of felt earthquakes.[3] Similar procedures have been applied to hydraulic fracturing operations in some states (e.g., Ohio): if earthquakes are detected, operations must be modified or suspended.[3]

Produced Water
Most oil- and gas-bearing rocks also contain water. When this is extracted along with oil and gas, it is called "produced water". For more on produced water reuse and disposal, see "Using Produced Water" in this series.

Earthquake Magnitude
Earthquake magnitude (M) describes the amount of energy released by an earthquake. For every two units of magnitude, the energy release is roughly 1000 times larger. M3 and larger can often be felt, and M5 can cause moderate damage. The largest earthquakes are all naturally occurring and can reach M9 or greater. Underground wastewater disposal has been linked to earthquakes as large as M5.8 in Oklahoma.[18]

References & More Resources

For a complete listing of references, see the "References" section of the full publication, *Petroleum and the Environment*, or visit the online version at: www.americangeosciences.org/critical-issues/petroleum-environment

Water Sources for Hydraulic Fracturing
Reducing the oil and gas industry's need for fresh water

Hydraulic Fracturing and Water Demand

Hydraulically fracturing a modern well can require millions of gallons of water for the initial fracturing process. This is a potential problem in arid regions with competing demands for fresh water (i.e. high water stress), such as Colorado and West Texas (see map). Today, advanced technologies allow the use of saline or brackish water (including groundwater and recycled oilfield water) for hydraulic fracturing, decreasing the demand for fresh water.

Using Brackish Groundwater

Roughly 14% of all the water used in the United States is too salty to drink. Most of this is seawater used for cooling at coastal power plants. Much of the rest is brackish (slightly salty[1]) groundwater, used in the oil and gas industry for hydraulic fracturing, for water injection to improve oil recovery, and in refineries.[2]

Using Produced Water

Oil and natural gas co-exist underground with varying amounts of water, so in some cases significant amounts of water may be extracted, or "produced", along with the oil and/or gas – nationally an average of 10 barrels of water are produced for each barrel of oil.[6] This produced water is often naturally salty, contains residual oil, and, for hydraulically fractured wells, may contain "flowback" water and chemicals from the original hydraulic fracturing fluid. Most produced water cannot be safely released into the surface environment, so over 90% is disposed of in deep underground injection wells.[6] Storing, treating, and re-using this water for hydraulic fracturing and other oilfield operations can help reduce the need for both disposal wells and fresh water.

Regional variations in the amount and composition of produced water, as well as differences in state regulations, affect the use of produced water in hydraulic fracturing and other oilfield applications such as waterflooding (injecting water into oil formations to help push out more oil):

- In the Bakken area of North Dakota only about 5% of the wells drilled in 2014 used produced water in their fracturing fluid. This is partly due to state regulations that prohibit storage of salty produced water in open-air pits,[7] and partly because the extreme salinity of produced water in this area makes treatment and reuse difficult and expensive.[8]

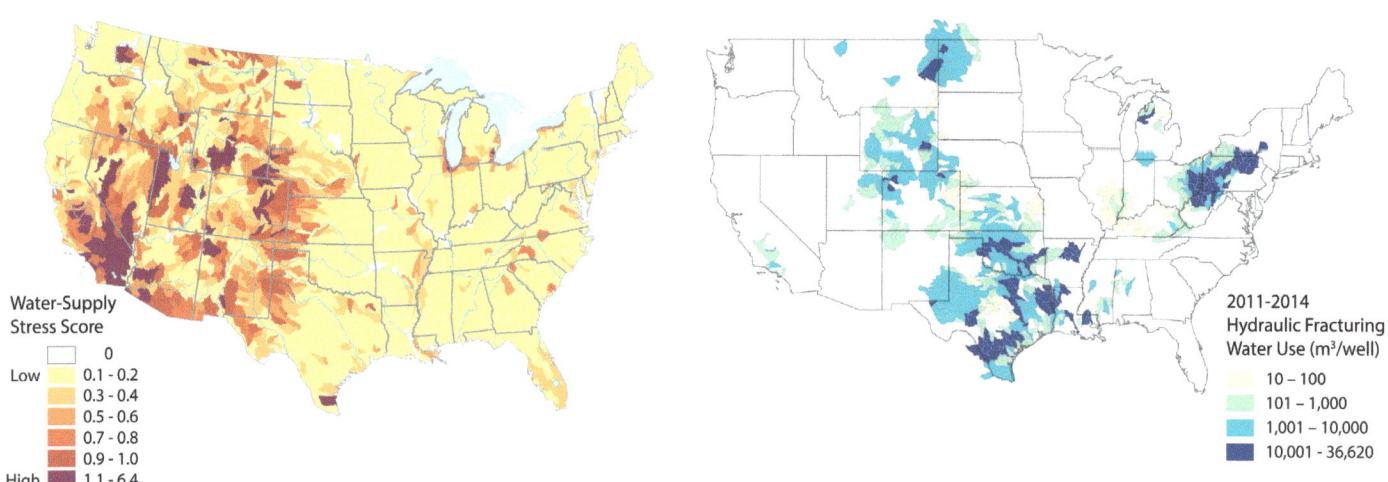

Left: Water stress in the United States[3] - in brown areas of the map, total water demand for all uses ranges from 40-80 percent of available supply.[4] Right: Hydraulic fracturing water use per well across the country.[5] Image credits: Left: Copyright, Union of Concerned Scientists, used with permission; Right: U.S. Geological Survey.

Petroleum and the Environment
Part 4: Water Sources for Hydraulic Fracturing

- The Marcellus shale in the northern Appalachians produces very little water compared to other major oil- and gas-producing regions.[9] Almost all of the produced water is reused in hydraulic fracturing operations, but the small amount of water produced compared to the amount used means that produced water can provide only a small fraction of the water needed for hydraulic fracturing in this area.[10]

- In Texas, reusing produced water is less attractive because the state has large amounts of brackish groundwater. It is cheaper to extract and use this groundwater than to store, transport, and treat produced water for reuse in hydraulic fracturing operations. Underground disposal of produced water is also relatively easy and inexpensive, and water ownership issues have discouraged transfer of produced water between operators.[11] As a result, companies in Texas are more likely to dispose of produced water than to treat, transport, and reuse it.

- The Eagle Ford shale in semi-arid South Texas has been a significant source of both oil and gas in the hydraulic fracturing boom. In this area, water needs for hydraulic fracturing are high, but less than 5% of this need is met by reusing produced and flowback water.[12] In 2013, hydraulic fracturing in the Eagle Ford was responsible for roughly 16% of total water consumption in the area. Much of this water was initially drawn from fresh groundwater supplies: between 2009 and 2013, one area saw groundwater levels drop by up to 200 feet due largely to extraction of fresh groundwater for hydraulic fracturing. Some operators here have switched to using abundant, non-potable brackish groundwater for hydraulic fracturing.[12]

Disclosing the Composition of Hydraulic Fracturing Fluids

FracFocus, an online chemical disclosure registry lists the chemicals used in hydraulic fracturing operations for over 124,000 wells (although some proprietary ingredients are undefined).[13] The registry is searchable by location, operator, or chemical.

In semi-arid West Texas, companies have long used brackish groundwater and reused produced water for both conventional and hydraulically fractured wells.[14] The abundance of brackish groundwater in Texas has also given rise to large-scale desalination operations to produce fresh water for industrial and municipal use.[15,16]

Alternative Hydraulic Fracturing Fluids

A variety of alternative fluids are being developed for hydraulic fracturing:

- To decrease environmental impacts caused by leaks and spills, less harmful additives are being developed and fewer additives are now added.

- To decrease water use, alternatives include fluids foamed with nitrogen or carbon dioxide, which can be used in low-pressure oilfields,[17] and high-pressure carbon dioxide. Some of these alternatives require specialized equipment and may be more expensive than standard hydraulic fracturing fluids.

References & More Resources

For a complete listing of references, see the "References" section of the full publication, *Petroleum and the Environment*, or visit the online version at: www.americangeosciences.org/critical-issues/petroleum-environment

American Water Works Association (2013). Water and Hydraulic Fracturing. http://www.awwa.org/portals/0/files/legreg/documents/awwafrackingreport.pdf

Gallegos, T.J. et al. (2015). Hydraulic fracturing water use variability in the United States and potential environmental implications. Water Resources Research, 51(7), 5839-5845. https://agupubs.onlinelibrary.wiley.com/doi/full/10.1002/2015WR017278

U.S. Department of Energy – The Energy-Water Nexus: Challenges and Opportunities. https://energy.gov/under-secretary-science-and-energy/downloads/water-energy-nexus-challenges-and-opportunities

Petroleum and the Environment
Part 5

Using Produced Water
Recycling oilfield water in the oil and gas industry and beyond

Opportunities and Concerns in Using Produced Water

Produced water is natural groundwater that is extracted along with oil and gas. It is commonly salty and mixed with oil residues,[1] so it must be either disposed of or treated and reused. About 2.5 billion gallons of produced water are extracted every day from all types of oil and gas wells, including coalbed methane wells.[2] If treated appropriately, produced water may be a valuable water source for agriculture, industry, and energy production,[3,4] but currently less than one percent of produced water is reused outside the oilfield.[5] This is due both to the high cost of treatment and to public concerns over the environmental and human safety of produced water.

Chemistry and Quantity of Produced Water Determine its Fate

Produced water reuse depends on the water's chemistry and quantity, the cost of treatment, transportation, and storage, and state and federal regulations. Produced water can contain varying amounts of dissolved oil and gas, toxic chemicals such as benzene, naturally occurring radioactive material, and dissolved solids or salts. The chemistry and quantity of produced water vary greatly from place to place and may also vary over the lifetime of a single well.

The salinity of produced water, a major constraint on reuse, ranges from nearly fresh (≤1%) to about 50% – 15 times saltier than seawater. Most produced water is currently disposed of in deep underground injection wells,[2] so access to disposal wells may also influence reuse decisions: in 2015, Texas had about 8,100 active oilfield wastewater disposal wells,[6] but in 2017, Pennsylvania had only 11 wells permitted for this purpose.[7] Under the Safe Drinking Water Act, underground injection of oil- and gas-related fluids for disposal or enhanced oil recovery is regulated by the EPA for nine states (including two oil-producing states: Pennsylvania and Virginia) or individually by the 41 states and two tribes to which the EPA has delegated authority.[8]

Produced Water Treatment

Before being reused, produced water must be treated to remove oil residues, salts, suspended solids, and other chemicals. The level of treatment required depends on what the water will be used for. When reusing produced water for new hydraulic fracturing operations, often relatively little treatment is required.[5] For irrigation or groundwater recharge, most of the salt must be removed. The more treatment required, the more expensive the process; it is possible to produce distilled water from produced water, but this would be very expensive. New technologies are continuously being developed and refined; for example, methods are being developed to remove boron, which is toxic to plants and can degrade some gels used for hydraulic fracturing.[5]

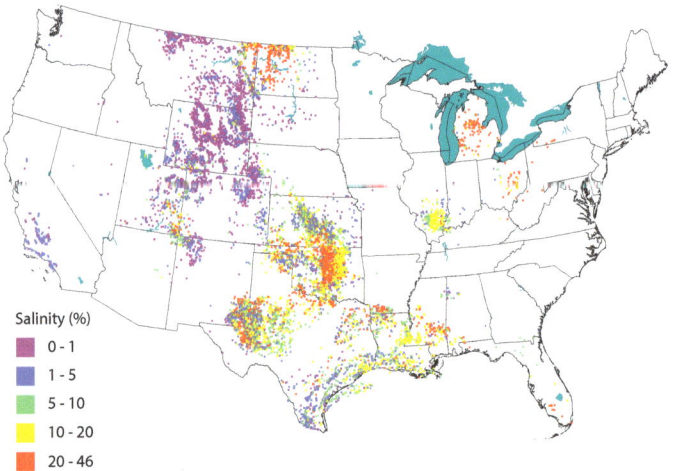

Salinity (total dissolved solids) of produced waters in the United States. The composition of produced water depends strongly on where it comes from and influences options for treatment and reuse. For comparison, seawater has a salinity of 3.5%. Image credit: Tracey Mercier, U.S. Geological Survey.

> ### Flowback Water
> In addition to naturally occurring produced water, some of the water that comes up a hydraulically fractured oil or gas well is flowback of the previously injected hydraulic fracturing fluid.

Petroleum and the Environment
Part 5: Using Produced Water

Reuse in the Oil and Gas Industry

- By far the most common application for reusing produced water is for injection into oil-producing rock formations to enhance oil production. Nationally, about 45% of all produced water is reused for this purpose.[2]

- In the Marcellus shale of Pennsylvania, West Virginia, and surrounding states, high water demand for hydraulic fracturing but low produced water volumes mean that almost all produced water is reused for hydraulic fracturing. Produced water only provides about 15% of the water needed to fracture new wells; the rest of the water comes largely from fresh surface waterbodies.[9]

- In the Permian basin of West Texas, companies that operate large numbers of wells reuse millions of gallons of produced water. Centralized treatment and storage facilities served by a network of pipelines help to reduce truck traffic, thus decreasing associated emissions and road damage.[2]

Reuse Beyond the Oilfield

Depending on the level of treatment, produced water may be used for many purposes. Some recent examples include:

- Thousands of acres in the Powder River Basin (Wyoming and Montana) are irrigated using treated produced water from coalbed methane wells to restore overgrazed range land or produce livestock forage. Some treated coalbed methane water is used to provide drinking water for livestock and wildlife.[10]

- In Wellington, Colorado, treated produced water is used in an aquifer storage and recovery project to maintain groundwater supplies in the region.[11]

- In California, low-salinity produced water is reused for enhanced oil recovery, groundwater recharge, and agriculture. Of the 80 billion gallons of oilfield water produced in California in 2013, about 20% (entirely from wells that are not hydraulically fractured) was treated and used for irrigation of crops for human consumption. State regulations specifically prohibit the use of water from hydraulically fractured wells to irrigate food crops.[12]

- Some states allow the spreading of highly saline produced water on roads for winter ice control.[13,14] The EPA recommends against this practice because the water may contain other pollutants in addition to salt.[15]

A Complicated Legal Framework

If produced water has economic value (as a water or mineral source rather than a waste product), its reuse is complicated by water rights laws and regulations, which vary from state to state. Another legal concern for production companies is liability for the reused water, which, depending on the level of treatment, may be unsuitable for certain uses.[5]

Mining Resources from Produced Water

Treatment of produced water typically involves removing a wide variety of dissolved salts and other compounds to improve the quality of the water. Some of these compounds may themselves have value; for example, several Oklahoma operations extract and sell iodine from highly saline produced water.[16] Other potentially valuable elements, such as lithium, also occur in produced water, but extraction of most elements is not currently economical.

References & More Resources

For a complete listing of references, see the "References" section of the full publication, *Petroleum and the Environment*, or visit the online version at: www.americangeosciences.org/critical-issues/petroleum-environment

U.S. Geological Survey – National Produced Waters Geochemical Database. https://energy.usgs.gov/EnvironmentalAspects/EnvironmentalAspectsofEnergyProductionandUse/ProducedWaters.aspx#3822349-data

National Academies of Science, Engineering, and Medicine (2017). Flowback and Produced Waters: Opportunities and Challenges for Innovation: Proceedings of a Workshop. Washington DC: The National Academies Press. https://www.nap.edu/catalog/24620/flowback-and-produced-waters-opportunities-and-challenges-for-innovation-proceedings

Petroleum and the Environment
Part 6

Groundwater Protection In Oil and Gas Production
Identifying and mitigating contamination of groundwater from oil and gas activity

Introduction

The United States relies on groundwater for roughly 25% of its fresh water.[1] This groundwater is found in porous, permeable rocks (aquifers) that often lie close to the Earth's surface – the deepest freshwater aquifers are found more than 6,000 feet underground,[2] but most are much shallower, from near the land surface to a few hundred feet below the surface.[3] In contrast, many of the largest oil and gas deposits are deeply buried many thousands of feet below the Earth's surface. As a result, oil and gas production involves drilling through aquifers to access the oil and gas farther below. Groundwater protection techniques have long played a crucial role in protecting environmental and human health during oil and gas production. More recently, oil and gas operations, particularly hydraulic fracturing, have raised concerns about the potential for aquifers to be contaminated by methane, produced water, or hydraulic fracturing fluid. This has led to more extensive research into the mechanisms, likelihood, and prevention of groundwater contamination.

Potential Contaminants

Hydraulic fracturing chemicals – hydraulic fracturing fluid is roughly 99% water. The remaining 1% typically consists of 3 to 12 chemical additives that improve the effectiveness of the fluid during hydraulic fracturing operations.[5] Some of these additives are toxic, and because a single hydraulically fractured well may use several million gallons of fluid, even a small leak may pose a risk to local groundwater supplies. Compositions of hydraulic fracturing fluids for most recent wells are listed in the FracFocus chemical disclosure registry,[6] although some components are kept confidential except in an emergency – this confidentiality exists to protect companies' investments in developing more effective fluids.

Drilling fluids – used to lubricate drills, remove rock chips, and maintain pressure in the well during drilling, usually consist mostly of mud and water with smaller amounts of minerals and chemicals that change the physical properties of the fluid to improve its function. Most of the substances in drilling fluids are not harmful to humans, and states typically impose restrictions on additives that can be used when drilling through freshwater aquifers. However, there have been rare cases in which unauthorized use of drilling fluid additives while drilling through freshwater aquifers has been found to contaminate groundwater.[7]

Methane – methane is a naturally occurring, flammable gas that is non-toxic but explosive at high concentrations. It is also a potent greenhouse gas. Methane is the main component of natural gas but it is also produced by microbes in sediments and wetlands, so methane found in groundwater may come

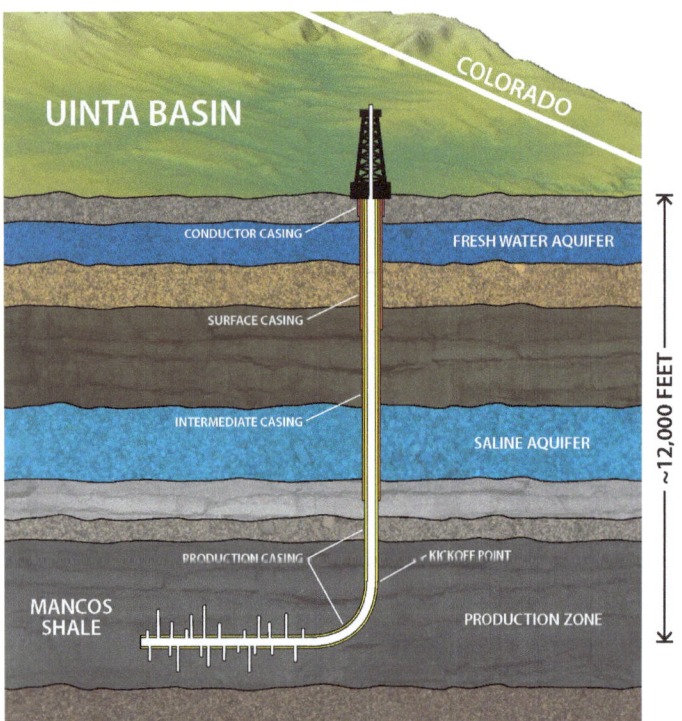

Schematic cross-section of a hydraulically fractured gas well in the Uinta Basin, Utah. The spatial relationships between the producing shale and overlying aquifers shown here are common for many, but not all, shales: operators must drill through fresh water aquifers to access the shales far beneath. Steel "casing" pipes are used to prevent the transfer of fluids between the well and the aquifer being drilled through. Hydraulically generated fractures in the shale may extend upwards for hundreds to more than 1,000 feet, but in most cases this still leaves thousands of feet of undisturbed rock between the hydraulic fractures and fresh groundwater. Image credit: Courtesy of the Utah Geological Survey.[4]

Petroleum and the Environment
Part 6: Groundwater Protection In Oil and Gas Production

from nearby oil and gas operations, natural microbes, or both.[8] Methane can also enter groundwater due to natural leakage from coal or gas-rich rocks. Determining the contributions of these different methane sources is a crucial step in assessing the environmental impacts of oil and gas production.

Oil and **produced water** – any fluid that can enter a well also has the potential to leak out of it if the well is compromised. This may include natural gas (see above), oil, or the often salty water that coexists with oil and gas in many rocks (called "produced water" when it is extracted along with oil and gas). Oil and saltwater leaks are rare but are of greater risk in improperly abandoned or "orphaned" wells than in active ones.[9]

Potential Mechanisms for Groundwater Contamination

There are three main ways in which oil and gas wells may contaminate groundwater: (1) if a well leaks (see "Preventing Well Leaks", below), (2) if oil or other fluids are spilled at the surface (see "Spills in Oil and Natural Gas Fields" in this series), or (3) if a hydraulic fracturing operation generates cracks in the overlying rocks it is theoretically possible for fluids and/or gas to move up through the rocks and into an aquifer.

Do hydraulically generated fractures create pathways for groundwater contamination?

The horizontal and vertical extents of hydraulically generated fractures are often determined during the fracturing process. Studies of thousands of hydraulic fracturing operations in the Barnett Shale (Texas) and the Marcellus Shale (Ohio, Pennsylvania, and West Virginia) have found that hydraulic fracturing operations took place more than 3,000 feet below any aquifers, and that the fractures generated during these operations generally extended upwards for only a few hundred feet.[11] A few fractures extended more than 1,000 feet, but in all cases there were still thousands of feet between the maximum extent of the fractures and the freshwater aquifers. Hydraulic fracturing for oil and gas in most areas therefore does not appear to generate fractures that allow for the migration of hazardous chemicals into freshwater aquifers. However, developing a complete understanding of these processes is an ongoing endeavor, and a procedure that is safe in one area or at one well may present risks in another.

Identifying Causes of Groundwater Contamination

Are hydraulic fracturing and/or leaking wells responsible for the contamination (primarily by methane) observed in some aquifers? This question has been extensively studied, debated, and litigated in a number of major oil- and gas-producing areas,[12] including Pavillion, Wyoming;[13] Dimock, Pennsylvania;[14] and Parker County, Texas.[15] In many cases, water quality samples were not collected prior to hydraulic fracturing, making it difficult to establish whether or not the aquifer contamination is related to hydraulic fracturing, earlier oil and gas production, or natural seepage. Most states now require (and many companies choose to conduct) pre-drilling testing of domestic and public water supplies so that the source of any future contamination can be more reliably identified.

Preventing Well Leaks

Oil and gas wells are constructed with multiple steel pipes ("casings") and cement barriers to prevent leaks of oil, water, or gas into aquifers. Casing and cement also prevent contamination

One of the main ways in which groundwater contamination may result from oil and gas activity is if fluids spilled on the surface seep down into groundwater. Above: the area affected by the 1979 pipeline spill in Bemidji, Minnesota, is now managed by the U.S. Geological Survey (USGS), which conducts research into the long-term effects of major spills. In this picture, USGS scientists are collecting a groundwater sample from a well to monitor how the different components of crude oil that has contaminated groundwater are broken down over time by biological activity. Image credit: U.S. Geological Survey.[10]

Petroleum and the Environment
Part 6: Groundwater Protection In Oil and Gas Production

of the well itself by material from the surrounding rock formations. An oil or gas well can leak if the steel casing or cement are damaged or poorly constructed. This is true of all wells, old and new, whether or not they are hydraulically fractured. The U.S. Environmental Protection Agency (EPA) has found that the hydraulic fracturing process itself is not a major contributor to well leaks from oil and gas operations.[12] More recent analyses have supported this finding, while also noting that surface spills are more common due to the large amounts of various different fluids being handled at and near the well site[16] (see "Spills in Oil and Natural Gas Fields" in this series for more information).

Federal or state regulators specify the depth of casing and cement layers based on local geology, depth of any freshwater aquifers, and contamination risk. Some oil- and gas-producing states, such as Pennsylvania, have updated their well casing and cementing regulations since 2008 to reduce the risk of leaks into aquifers.[17]

Examples of Methane Leaks from Wells

Although methane is often naturally present in aquifers, leaks from oil and gas wells are possible and may include other, more hazardous chemicals. A variety of attempts have been made to determine the number and causes of these leaks:

- The Denver-Julesburg area north of Denver, Colorado, contains about 54,000 oil and gas wells. Data from over 900 water wells in the area found dissolved methane in almost 600 of them. In the vast majority of cases, this methane occurred naturally in the groundwater due to microbial activity. However, in roughly 40 water wells measured between 2001 and 2014, the chemistry of the methane indicated that it came from leaking, often older, oil and gas wells.[18] The boom in horizontal drilling and hydraulic fracturing began in 2010, but from 2010 to 2014 no increase in contamination rates was observed, suggesting that these practices did not increase well leaks.

- In the Marcellus shale area of Pennsylvania, West Virginia, and surrounding states, the picture is similar. Most methane found in groundwater is naturally produced by microbes, but in rare cases leaking wells can contaminate groundwater (see above).[19]

- The EPA's 2016 assessment of the impacts of hydraulic fracturing on water resources found that hydraulic fracturing can increase the risk of leaks in poorly constructed wells. In 2007, improper cementing of a new well in Ohio and subsequent hydraulic fracturing allowed gas to enter a freshwater aquifer and then migrate into a house through a domestic water well. Accumulation of this gas caused an explosion that damaged the house.[20]

- Old, abandoned wells may be less well constructed and may not have been properly plugged before being abandoned. Leaks from these wells may be more likely – see "Abandoned Wells" in this series for more information.

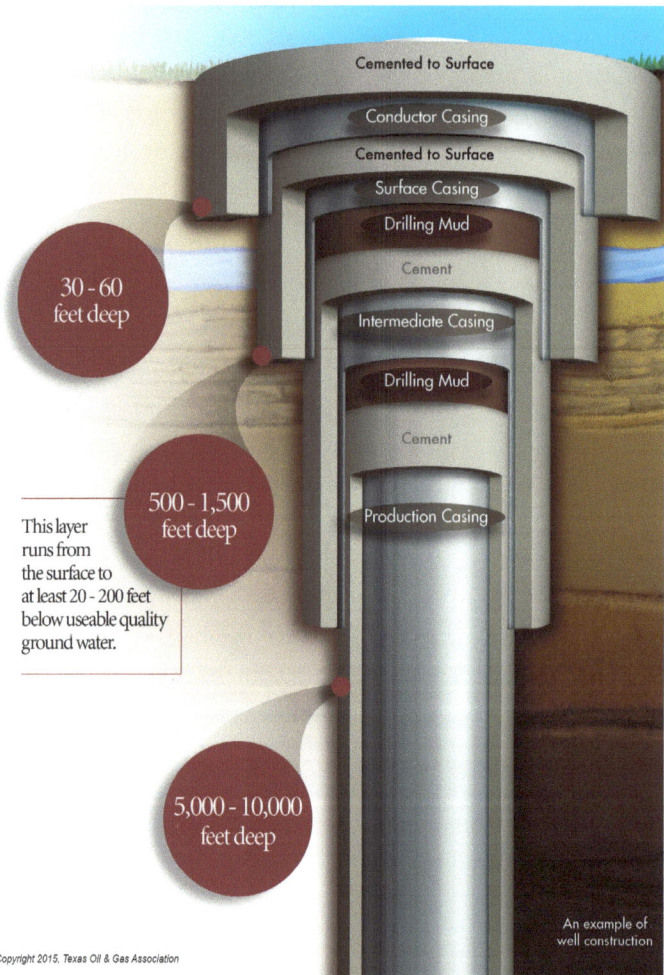

Several layers of steel casing and cement are used to prevent leaks out of or into an oil or gas well. Image not to scale. Image credit: Courtesy of Texas Oil and Gas Association.

Petroleum and the Environment
Part 6: Groundwater Protection In Oil and Gas Production

Naturally Occurring Methane in Aquifers

In oil- and gas-producing areas, the presence of methane in aquifers can cause significant local concern over the safety of oil and gas operations. In many cases, however, methane is naturally present in the local rocks and soil:

- In the Marcellus shale area of Pennsylvania, West Virginia, and parts of adjacent states, methane was commonly found in aquifers before any hydraulic fracturing took place in the area. Measurements taken before nearby hydraulic fracturing activities began found methane in 24% of 189 water wells in Pennsylvania.[21] Pre-drilling methane contamination in the Marcellus region is caused by both microbial activity and the slow, natural leaking from gas-rich rocks in the area.[22]

- A 2017 study by the U.S. Geological Survey found that in the main natural-gas-producing areas of Texas, Louisiana, and Arkansas, most methane found in water wells is produced by microbes, not leaking gas wells.[23] This was determined by studying the chemistry of the methane and associated gases, which can be used to distinguish microbial methane from natural gas.

References & More Resources

For a complete listing of references, see the "References" section of the full publication, *Petroleum and the Environment*, or visit the online version at: www.americangeosciences.org/critical-issues/petroleum-environment

Ground Water Protection Council (2017). State Oil and Natural Gas Regulations Designed to Protect Water Resources. Third Edition, November 2017. http://www.gwpc.org/sites/default/files/State%20Regulations%20Report%202017%20Final.pdf

National Ground Water Association (2013). Water Wells in Proximity to Natural Gas or Oil Development. NGWA Information Brief, updated September 2017. http://www.ngwa.org/Media-Center/briefs/Documents/Info-Brief-Hydraulic-Fracturing.pdf

U.S. Environmental Protection Agency (2016). Hydraulic Fracturing for Oil and Gas: Impacts from the Hydraulic Fracturing Water Cycle on Drinking Water Resources in the United States (Final Report). EPA/600/R-16/236F. https://cfpub.epa.gov/ncea/hfstudy/recordisplay.cfm?deid=332990

Petroleum and the Environment
Part 7

Abandoned Wells
What happens to oil and gas wells when they are no longer productive?

Introduction

In 2017, there were one million active oil and gas wells in the United States.[1] When a well reaches the end of its productive life, or if it fails to find economic quantities of oil or gas, the well operator is required by regulators to remove all equipment and plug the well to prevent leaks.[2] Usually, cement is pumped into the well to fill at least the top and bottom portions of the well and any parts where oil, gas, or water may leak into or out of the well. This generally prevents contamination of groundwater and leaks at the surface. State or federal regulators define specific plugging procedures depending on the local conditions and risks, and may monitor the plugging operation.

However, there are many cases in which wells are not properly plugged before being abandoned, especially if the well operator goes bankrupt, leaving its wells "orphaned".[3] This is more common when oil prices fall rapidly, making many wells uneconomical, as in the 1980s oil glut, the 2008 financial crisis, and the 2014 downturn.

In the late 1980s, the U.S. Environmental Protection Agency estimated that 200,000 of 1.2 million abandoned wells may not have been properly plugged.[4] Since then, tens of thousands of orphaned wells have been plugged by state and federal regulators, as well as some voluntary industry programs. These efforts are ongoing, and many orphaned wells have yet to be properly plugged. The exact number is not known: some 3.7 million wells have been drilled in the U.S. since 1859,[6] and their history is not always well documented. Older wells, especially those drilled before the 1950s, are particularly likely to have been improperly abandoned and poorly documented.

Risks to Groundwater, Air, and the Surface Environment

Orphaned wells are often abandoned without any plugging or cleanup, but even plugged wells may leak, especially those plugged in the past, when plugging procedures were less rigorous and used less durable materials. Unplugged or poorly plugged wells may affect:

- **Groundwater** – old wells may have degraded well casing or cement that can allow oil, gas, or salty water to leak into freshwater aquifers. An assessment of 185 groundwater contamination incidents in Ohio from 1983 to 2007 found 41 incidents caused by leakage from orphaned wells, compared to 113 incidents caused during drilling and production.[7]

- **Methane emissions** – a study of 138 abandoned wells in Wyoming, Colorado, Utah, and Ohio found that over 40% of unplugged wells leaked methane, compared to less than 1% of plugged wells. This study estimated that abandoned wells account for 2-4% of the methane emissions from oil and gas activity.[8]

- **The surface environment** – orphaned sites may still have old equipment, contaminated soil from small spills, and other waste at the surface. In some unplugged or poorly plugged wells, oil, gas, drilling mud, or salty water can rise up the well and spill at the ground surface or, in the case of offshore wells, into open water.[9]

Inoperable pumpjack at an abandoned well in Texas. Image credit: Steve Hillebrand, U.S. Fish and Wildlife Service.[5]

Petroleum and the Environment
Part 7: Abandoned Wells

Abandoned Wells and Hydraulic Fracturing

Hydraulic fracturing uses the high-pressure injection of fluids into oil- or gas-bearing rocks to fracture them and allow oil and/or gas to flow out. The increased pressure in the rocks during this process can push oil or salty water up nearby unidentified or improperly plugged abandoned wells. In one of the more extreme cases of this, the U.S. Environmental Protection Agency cited an abandoned well in Tioga county, Pennsylvania, that produced a 30-foot geyser of brine for more than a week as a result of hydraulic fracturing of a nearby well.[10] In addition to these fluids contaminating soil and potentially reaching groundwater, the unexpected pressure release caused by this fluid escape reduces the effectiveness of the hydraulic fracturing operation, so there are both environmental and economic incentives to identify and plug wells near a planned hydraulic fracturing operation.

Abandoned Well Plugging Campaigns

For several decades, states have increased enforcement of plugging and cleanup requirements. States generally require a performance bond or other financial assurance from the operator that a well will be plugged and the well site restored. However, bond amounts may not meet the plugging and cleanup expenses if an operator goes bankrupt.[11] Most states therefore collect fees or a production surcharge from operators specifically for remediation of orphaned wells and associated surface equipment.[12] For example, Pennsylvania adds an orphaned well surcharge to drilling permit application fees,[14] while Texas adds a 5/8-cent Oil Field Cleanup surcharge to the state's 4.6% oil production tax.[15] The Oklahoma Energy Resources Board remediates abandoned well sites using voluntary industry contributions amounting to 0.1% of oil and gas sales.[16]

Examples of Plugging Programs

- From 1984 to 2008, the Railroad Commission of Texas (RRC, the state's regulatory agency for oil and gas) plugged almost 35,000 orphaned wells, including offshore wells, at a cost of over $163 million.[17] In fiscal year 2017, the RRC plugged 918 orphaned wells at a cost of over $11.6 million.[18] As of December 2017, there were roughly 10,000 known orphaned wells in Texas that required plugging; the RRC aims to plug 1,500 of these in fiscal year 2018.[19]

- From 1989 to 2017, the Pennsylvania Well Plugging Program plugged over 3,000 orphaned wells.[20]

- The voluntary Oklahoma program mentioned above has cleaned up and restored 15,000 orphaned and abandoned well sites in Oklahoma since 1994 at a cost of almost $100 million.[21]

- California has plugged more than 1,350 orphan wells since 1977 at a cost of over $27 million. In 2016, many wells remained to be plugged, including about 900 in the city of Los Angeles. The California Department of Conservation is authorized to spend $1 million per year to remediate orphan wells.[22]

- The U.S. Bureau of Land Management (BLM) reclaimed 295 orphaned well sites in ten states from 1988 to 2009 at a cost of $3.8 million. As of 2010, BLM estimated that there were 144 orphan wells yet to be reclaimed in seven states.[23]

An abandoned well site in Oklahoma before (top) and after (bottom) being remediated. Image credit: Courtesy of the Oklahoma Energy Resources Board.[13]

References & More Resources

For a complete listing of references, see the "References" section of the full publication, *Petroleum and the Environment*, or visit the online version at: www.americangeosciences.org/critical-issues/petroleum-environment

Petroleum and the Environment
Part 8

What Determines the Location of a Well?
Geology, leasing, permitting, technology, economics, and the environment

Introduction

Oil- and gas-rich rocks are only found in certain parts of the United States, so most of the country has no oil or gas wells. Where oil and gas production is commercially viable, many factors determine the exact location of each well, including leasing, permitting, competing land uses, environmental protection, economics, and drilling technology. These factors are strongly interlinked: the best well location for environmental protection may be on land that the owner will not lease for drilling, may be more expensive to drill on, or may be too close to a school for the state to issue a drilling permit. The most profitable location may endanger a local waterway or encroach on important agricultural land. The final location for most wells is usually chosen to balance the requirements of federal, state, and local regulators, landowners, oil and gas operators, and the local community.

Finding the Right Rocks

Oil and gas cannot form in all rocks; significant quantities can only form from organic matter (the remains of living organisms) trapped in thick layers of sediments. While peat and coal are formed from plant matter, oil and gas are mostly formed from large accumulations of tiny plankton that lived and died in ancient lakes, seas, and oceans. As the organic-rich sediments accumulate over time, the deeper portions become buried. Once the sediments reach depths of 5,000 to 30,000 feet, heat and pressure are high enough to convert the organic matter to oil and gas. The buoyant oil and gas may rise toward the surface and become trapped by overlying impermeable rocks or may remain trapped in the rocks where they formed (typically shales). These processes have produced the large oil and gas fields in California, within and east of the Rocky Mountains, and in the Gulf of Mexico and the Appalachian Basin.

Within oil-rich areas, finer-scale variations in the geology of the area will produce regions of greater or lesser productivity, as shown on the next page for the Marcellus and Utica/Point Pleasant plays in the Appalachian Basin.[3]

Leasing: Permission from the Landowner

Mineral rights – the ownership of rocks, minerals, oil, and gas beneath an area of land – may belong to private individuals or local, state, or federal government. Before exploring for and producing oil and gas, operators must obtain a lease from the mineral rights owner. The mineral rights owner may be different

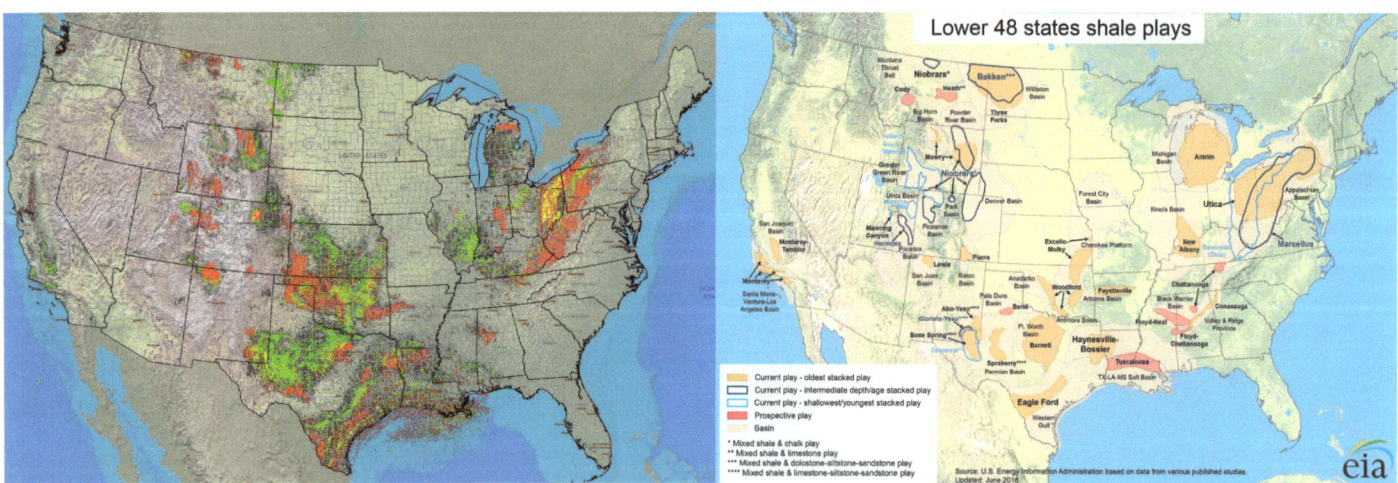

Left: Areas of historical oil (red), gas (green), or mixed (yellow) production in the contiguous United States as of 2005 (immediately prior to the shale boom). Right: Current (solid orange, plus blue and black outlines) and prospective (solid red) shales for oil/gas production, overlain on major sedimentary basins (tan), as of 2016. Image credits: Laura R.H. Biewick, U.S. Geological Survey;[1] U.S. Energy Information Administration.[2]

AGI Critical Issues Program: www.americangeosciences.org/critical-issues
Supported by the AAPG Foundation. © 2018 American Geosciences Institute

Petroleum and the Environment, Part 8/24
Written by E. Allison and B. Mandler for AGI, 2018

Petroleum and the Environment
Part 8: What Determines the Location of a Well?

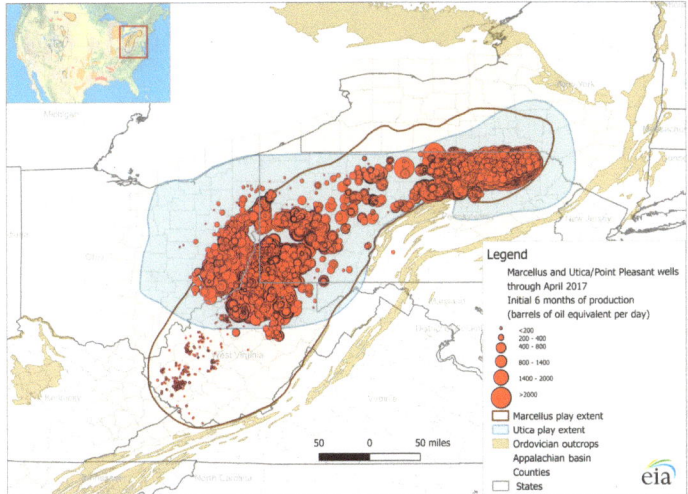

Map of wells in the Marcellus and Utica/Point Pleasant formations (Pennsylvania, Ohio, and West Virginia) through April 2017. Although potentially oil- and/or gas-bearing shales (brown outline and blue field) underlie almost the entire region, a wide variety of factors (discussed throughout the text) determine exactly where and in what concentration wells are drilled. Image credit: U.S. Energy Information Administration.[4]

from the surface landowner, and in some cases the owner of one mineral, such as coal, differs from the owner of oil and gas. These cases of "split estate," where there are different owners of surface and underground resources, result from the fact that mineral ownership can be sold or transferred like other property.

Obtaining a lease does not mean that a well will be drilled. Typically, an operator acquires a large number of leases in an area to give them the flexibility to drill wells in a range of locations based on the results of exploration and early drilling.

Private mineral owners can choose to lease or not to lease their land for drilling, and can negotiate the terms of the lease, including an up-front "bonus" payment of less than one hundred to as much several thousands of dollars per acre. The lease also sets the royalty payments (the proportion of the value of resources produced that will be paid to the mineral owner, often 12.5%). Lease terms may also include requirements to protect crops, livestock, or buildings, all of which may affect the location of the well.

The federal government restricts oil and gas activities in national parks, monuments, and areas where Congress or the President has suspended such activities. In other areas, the federal government leases public land for oil and gas development when it is deemed to be compatible with other public uses and the protection of wildlife, scenery, water, and land.[5] Federal leases are offered in regularly scheduled competitive lease sales. State lands may also be leased, usually in a competitive process. For example, the Marcellus and Utica shales underlie 1.5 million acres of Pennsylvania state forest land; in 2017, over 130,000 of these acres were under lease for shale gas production.[6] In addition, Native American tribes, individual Native American mineral owners, and Alaska Native Corporations may lease their lands for oil and gas development.[7]

Drilling Permits

Regardless of leasing agreements, operators must obtain permission from the state to drill a well on any land within that state, whether it is private or owned by the local, state, or federal government.[8] If the land is federally owned, federal approval is also required (see below). In some cases, the well will also require county or local government permission to drill. Although state drilling regulations vary widely, common regulations include:

- Restrictions on drilling in or near parks or historic sites

- Rules on how closely wells can be spaced, to prevent operators from extracting resources that belong to adjacent leases and mineral rights owners

- Minimum distances or setbacks between wells and homes, businesses, schools, roads or public areas – some local and county governments also set minimum distances within their jurisdictions

In some states, the application to drill a well may be open for public comments before being approved.

If the operator plans to hydraulically fracture a well, they may be subject to additional regulations such as taking groundwater measurements before drilling. This provides information about the pre-drilling composition of local groundwater, which allows potential groundwater contamination from hydraulic fracturing fluids to be identified.[9] As of 2018, hydraulic fracturing has been banned in two states with resources that could be produced using the technology (New York[10] and Maryland[11]) and one state with no known oil or gas resources (Vermont[12]).

Petroleum and the Environment
Part 8: What Determines the Location of a Well?

Drilling Restrictions on Federal Land

Drilling on public land is controlled by the federal government, which aims to balance a wide variety of land uses, including oil and gas exploration and production, livestock grazing, hunting and fishing, coal and mineral development, recreation, and natural or cultural conservation.[13] Laws passed by Congress and signed by the President – or executive orders signed by the President – can restrict or ban leasing and/or drilling in federally controlled areas. Recent and historical restrictions and bans of this nature have been implemented in national parks and monuments, wilderness areas, the Great Lakes, and offshore (see "Offshore Oil and Gas" in this series for more information on offshore drilling).

Oil and gas development of onshore federal land is largely overseen by the Bureau of Land Management (BLM). There are roughly 100,000 active wells in areas managed by BLM; between 2000 and 2016, on average 3,000 new wells were drilled each year.[13,14] The Forest Service, the National Park Service, the Army Corps of Engineers, the military, or the Bureau of Reclamation may impose additional restrictions on lands under their management. Regardless of land ownership, if operators intend to drill beneath a navigable waterway or add new material (e.g., for roads or well pads) that might affect a waterway or wetland, they must first obtain permission from the U.S. Army Corps of Engineers.[15]

The National Environmental Policy Act of 1970 (NEPA) requires federal agencies to assess the environmental effects of their proposed actions, including oil and gas leasing or drilling on federal land.[16] For NEPA compliance and other requirements, BLM develops regional, long-term land-use plans, called Resource Management Plans, with input from other government agencies, individuals, organizations, and local governments (see "The Pinedale Gas Field, Wyoming" in this series for example elements of a Resource Management Plan).[17]

Drilling Technology

Early oil and gas wells were drilled straight downward, meaning that oil and gas resources could only be extracted if a well site could be installed directly above them. Over time, drilling technology continuously advanced, and by the first half of the 20th century, wells could be drilled at an angle, allowing the location of the wellhead to be placed away from sensitive areas or competing land uses. More recently, advances in horizontal drilling have allowed operators to drill horizontally underground for up to several miles.[20] This has the potential to provide increased flexibility in choosing the surface location of drill sites based on other factors, such as environmental protection (see sidebox).

Protecting Forests while Producing Energy in Appalachia

Forests and streams in Appalachia support diverse plant and animal populations that may be threatened by habitat fragmentation caused by oil and gas activity. Environmental organizations in partnership with industry and academia are working to reduce the surface impact of oil and gas operations in this region by optimizing the placement of drilling and production sites, roads, and pipelines. For example:

- The Landscape Environmental Energy Planning tool (LEEP)[18] was developed by The Nature Conservancy (TNC) in collaboration with the University of Tennessee at Knoxville, the Cadmus Group, and industry advisors. LEEP is an interactive, web-based GIS program that industry planners can use to assess the relative costs and environmental impacts of different configurations for wells, access roads, and gathering pipelines.

- A 2015 workshop hosted by TNC and Carnegie Mellon University brought together a group of over 70 organizations to develop siting recommendations for energy infrastructure to protect Appalachian biodiversity. The publication, "Advancing the Next Generation of Environmental Practices for Shale Development,"[19] was a collaborative effort between the energy industry, non-governmental organizations, academic institutions, and federal, state, and local government.

Enhanced Recovery Wells and Pooling

Many advanced techniques for oil and gas production require operations to span more than one lease:

- Enhanced oil recovery techniques such as steamflooding, waterflooding, and carbon dioxide injection require

Petroleum and the Environment
Part 8: What Determines the Location of a Well?

an operator to drill injection wells some distance from producing wells.[22]

- Since the early 2000s, the proliferation of horizontal wells extending a mile or more away from the vertical portion of the well has required operators to work with mineral rights owners in leases adjacent to those where the producing well sites are located.

"Pooling" refers to the combining of leases and sharing of operational costs and production revenues within those leases by all parties involved. Mineral owners adjacent to an area leased for oil and gas development may be legally forced to lease their subsurface minerals if theirs are necessary for development of the first lease – this is referred to as "forced pooling". All mineral rights owners in the pool share in the costs and revenues, even if there are no producing wells on their land.[23]

Economic Considerations

For oil and gas operations to be financially viable, the expected cost to explore for, drill, and extract the resource must be less than the value of oil and gas expected to be produced over the life of the well. National and global demand and price for oil and gas have major implications for the number and location of new wells that are drilled. For example, declining natural gas prices from 2008 to mid-2016 led operators to focus on more oil-rich areas. The drop in the price of oil in late 2014 reduced the total number of new wells being drilled and led the industry to focus operations in areas with highly productive wells and lower operating costs. Even with decreased drilling costs and improved drilling efficiency,[24] the number of new wells plummeted. For example, in Texas, the number of new wells drilled per year dropped from over 27,000 in 2014 to less than 9,000 in 2016.[25]

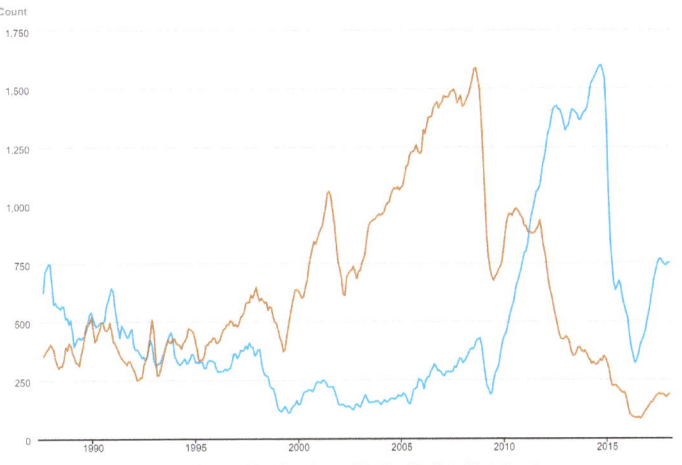

U.S. oil and natural gas drilling activity by number of drilling rigs in operation, 1988-2018. Natural gas drilling boomed with the use of horizontal drilling and hydraulic fracturing in the early 2000s, then fell in response to the 2008 recession and continued to fall due to excess gas production. Drilling in oil-rich shale areas picked up after the 2008 recession but fell with low oil prices in 2014-2015. Image credit: U.S. Energy Information Administration.[21]

References & More Resources

For a complete listing of references, see the "References" section of the full publication, *Petroleum and the Environment*, or visit the online version at: www.americangeosciences.org/critical-issues/petroleum-environment

Biewick, L.R.H. (2008). Areas of Historical Oil and Gas Exploration and Production in the United States. U.S. Geological Survey Digital Data Series DDS-69-Q. https://pubs.usgs.gov/dds/dds-069/dds-069-q/text/pdfmaps.htm

The Nature Conservancy – LEEP: The Nature Conservancy's Appalachian Shale Siting Tool. https://www.nature.org/ourinitiatives/regions/northamerica/areas/centralappalachians/leep-summary.pdf

The Nature Conservancy and Carnegie Mellon University (2016). Advancing the Next Generation of Environmental Practices for Shale Development: Workshop Deliberations and Recommendations. May 27-29, 2015. Pittsburgh, PA. https://www.cmu.edu/energy/documents/Shale_Workshop_Deliberations_and_Recommendations_Final.pdf

Pennsylvania Department of Conservation and Natural Resources (2017). Natural Gas Development and State Forests: Shale Gas Leasing Statistical Summary, May 2017. http://www.docs.dcnr.pa.gov/cs/groups/public/documents/document/dcnr_20029363.pdf

Petroleum and the Environment
Part 9

Land Use in the Oil and Gas Industry
How is land used? How do technologies and regulations minimize land use?

Introduction
All energy production requires land. Reducing the land-use "footprint" of the energy industry is an important part of limiting environmental impacts while meeting our energy needs. Advanced exploration technologies such as 3D seismic imaging, and drilling technologies such as horizontal and slanted wells, reduce the amount of land disturbed for a given amount of oil or gas produced.

The Size Of Well Sites
Drilling and hydraulically fracturing a well requires several acres around the well for the drilling rig, drill pipe storage, trailers for equipment and staff, pump trucks, data vans, and pits or tanks for water and waste storage. Once drilling is finished and the well is producing oil or gas, much of the drill site can be reclaimed. State and federal regulators require and oversee these reclamation efforts.[1,2,3,4] The size of a well site, or "pad", will depend on many factors, including location, land use restrictions, and the type and number of wells being drilled from the site. To take just one example, the total land footprint of a typical well site in Pennsylvania's portion of the Marcellus Shale is five to eight acres – this includes the land taken up by water impoundments for hydraulic fracturing, access roads, and other equipment.[5] For more information on how the land footprint of a well site

Marcellus Shale Energy and Environment Laboratory drill site near Morgantown, West Virginia. Image credit: Photo courtesy of Northeast Natural Energy.

may change over time, see "The Pinedale Gas Field, Wyoming" in this series.

Advances in Drilling Technology: Fewer, Smaller Well Sites
Modern drilling technology allows a drill to be guided horizontally underground for up to several miles.[7] This allows operators to avoid sensitive environments and drill multiple wells in different directions from a single site. As drilling technology continues to improve, longer horizontal wells will continue to reduce the number of sites needed.

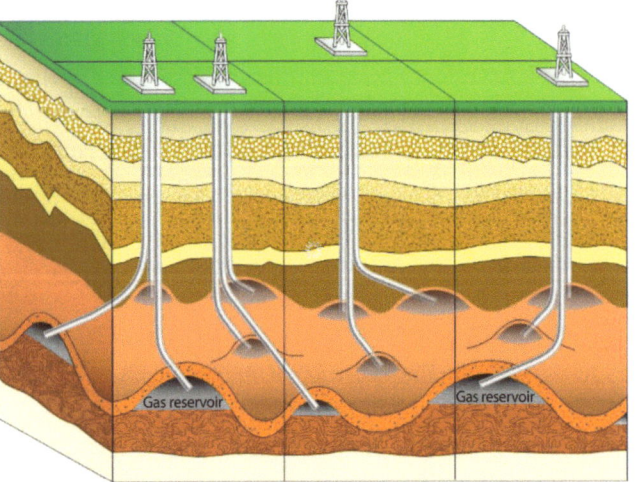
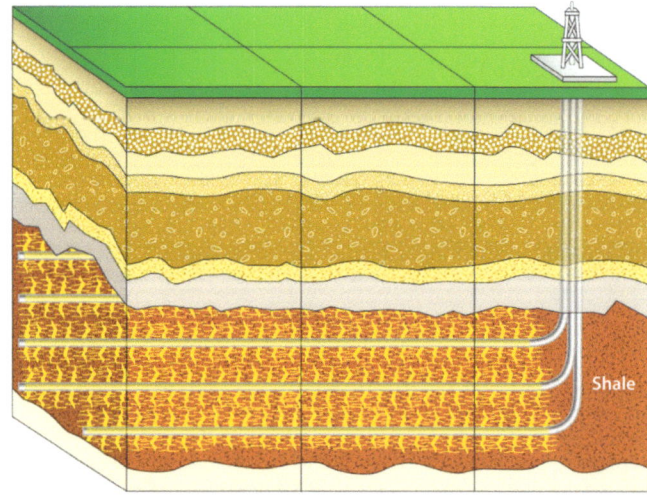

Schematics showing how multiple wells can be drilled in different angles and directions from a single site, reducing land use. Left: directional drilling to reach multiple lens-like gas reservoirs. Right: horizontal drilling and hydraulic fracturing of shale. Image credit: Government Accountability Office.[6]

AGI Critical Issues Program: www.americangeosciences.org/critical-issues
Supported by the AAPG Foundation. © 2018 American Geosciences Institute

Petroleum and the Environment, Part 9/24
Written by E. Allison and B. Mandler for AGI, 2018

Petroleum and the Environment
Part 9: Land Use in the Oil and Gas Industry

Advances in Seismic Imaging: Fewer Unproductive Wells

Exploration for oil and gas requires a detailed image of the rocks below the Earth's surface. Seismic imaging – like an ultrasound for the Earth – helps geoscientists to identify rock layers that are most likely to contain oil and gas, reducing the number of unproductive wells that are drilled. This in turn improves the land-use efficiency of oil and gas production. In the early 1970s, only 60% of wells drilled in the U.S. successfully hit an economically viable oil or gas deposit. By 2005, this had improved to 90%, due largely to improvements in 2-D and then 3-D seismic imaging, as well as other tools for monitoring and analysis.[8] This improvement has been seen not only in newly explored areas but also in existing oilfields, where a more detailed image helps operators to drill new wells in the most productive parts of the field.

Changes in a Well's Footprint Over Time

The initial development of an oil- or gas-producing area involves acquiring seismic information and drilling wells to identify the total size of the oil or gas reservoir and define the areas likely to be most productive. This process largely uses single-well drilling sites with a high land-use footprint.

Once this initial effort is completed, the relative land disturbance decreases as multiple wells are drilled from each site, unproductive areas are avoided, companies use centralized equipment and fluid handling facilities, and pipelines replace trucks for moving water, oil, and gas.

Once a well starts producing oil and/or gas, the drilling equipment is removed and much of the site is remediated. Supporting infrastructure, such as tanks, pipelines, and access roads, remain in place as long as the well is active, which may be many decades. Automated data collection can reduce the number of trips needed and the resulting land disturbance.

When a well is decommissioned, states require the operator to remove all surface equipment and plug the well with cement. State regulatory agencies maintain plugging records, which are commonly publicly available.[9]

Unproductive wells may sit idle for years, and are sometimes abandoned without being adequately sealed, especially when companies go out of business. Improperly abandoned wells pose particular environmental risks – see "Abandoned Wells" in this series for more details.

Roads, Plants, and Pipelines

The land footprint of oil and gas extends far beyond the wells:

- New production areas require new access roads, pipelines, and other infrastructure.
- Natural gas must be purified and crude oil must be refined before they are used. As of 2014, there were over 550 gas processing plants[10] and over 140 oil refineries[11] in the United States.
- Processed/refined products must be distributed by truck, rail, pipeline, or boat. Each mode of transport has different impacts on land- and energy-use – see "Transportation of Oil, Gas, and Refined Products" in this series for more on this topic.
- Other land uses include facilities for the treatment and disposal of wastewater; natural gas power plants; and gasoline service stations.

The modern energy system is vast. In areas of intense production, land dedicated to oil and gas activity can be very high as a proportion of total land use. Overall, however, the high energy density of oil and gas results in relatively little land disturbance for a large amount of energy production.[12] As the national and global energy sectors evolve over time, minimizing the land footprint of all energy technologies remains a key issue in attempts to balance energy production and environmental protection.

References & More Resources

For a complete listing of references, see the "References" section of the full publication, *Petroleum and the Environment*, or visit the online version at: www.americangeosciences.org/critical-issues/petroleum-environment

Jones, N.F. et al. (2015). The Energy Footprint: How Oil, Natural Gas, and Wind Energy Affect Land for Biodiversity and the Flow of Ecosystem Services. BioScience, 65(3), 290-301. https://academic.oup.com/bioscience/article/65/3/290/236920

Petroleum and the Environment
Part 10

The Pinedale Gas Field, Wyoming
A case study of changes in land use during exploration and production

Introduction
The Pinedale field is the sixth-largest gas field in the United States.[1] The core development area covers about 70 square miles in a sparsely populated area of southwest Wyoming, 70-100 miles north of Rock Springs.[2] In 2015, the Pinedale field produced 4 million barrels of gas condensate and 436 billion cubic feet of natural gas,[3] making it the largest natural gas-producing field in Wyoming (for comparison, the vast Marcellus Shale in the Appalachian basin produced 6 trillion cubic feet of natural gas in 2015[4]). Through 2012, the field had produced 3.9 trillion cubic feet of natural gas – only 10% of its potentially recoverable gas reserves – so operations may continue for many years.[5]

Field operations impact nearby small towns and large populations of sage grouse (under special protections[6]), pygmy rabbits (endangered in some of their range[7]), and pronghorn.[1] The field is also an important winter range for thousands of mule deer.[8] Since the field started production in the late 1990s, improvements in drilling, land use, air emissions, and water handling practices have reduced the physical, societal, and environmental impacts of drilling and production in the Pinedale field.

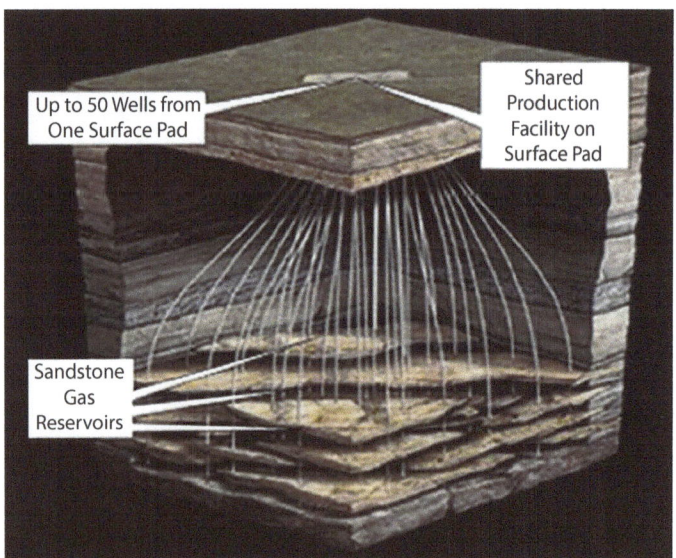

Schematic showing how curved wells allow a large area of gas-producing rock to be accessed from a single well site. Image credit: AAPG Wiki.[5]

Although Pinedale provides a good example of improvements in low-impact gas production, continued improvements will be important to the long-term preservation of this region throughout the field's production, eventual closure, and restoration to its pre-drilling state.

Directional Drilling from Multi-Well Pads
The gas-producing zone of rocks in the Pinedale field is almost 6,000 feet thick and made of many lens-shaped layers of sandstone. Over 2,500 curved wells have been drilled to tap into the producing zone, 8,500-14,000 feet underground, at distances thousands of feet horizontally from the drilling site. These wells are hydraulically fractured to allow gas to flow out of the otherwise impermeable sandstone.

Federal Management Plan
Over 80% of the field area is federal land overseen by the U.S. Bureau of Land Management (BLM), which must approve an operator's plan of operations and reclamation before drilling. Commercial development of the field was enabled by the use of hydraulic fracturing techniques that were applied starting in the 1990s. Early operations were governed by a 1988 Resource Management Plan/Environmental Impact Statement[9,10] developed by BLM with input from the public and other government agencies. The Resource Management Plan/Environmental Impact Statement was updated in 2008[11] to require or promote environmentally beneficial changes while allowing additional drilling.

Innovations that Reduce the Land Use and Environmental Impact of Operations[12]
- The 2008 Resource Management Plan (RMP) allows for the construction of multi-well sites concentrated in designated development areas, located away from streams and nesting, calving, and winter grazing areas. This allows drilling and other operations throughout the year in the designated development areas while reducing habitat fragmentation and leaving 92% of the Pinedale area undisturbed. Well sites used in early field development

Petroleum and the Environment
Part 10: The Pinedale Gas Field, Wyoming

Aerial photos of a well site in the Pinedale Field. Left: Well site during drilling in 2009. Right: Same site in 2015 – the site contains 17 wells, its area has been reduced from 15.3 to 6.04 acres, and the reclaimed 9.26 acres are being revegetated. Image credit: Courtesy of Timothy Zebulske, BLM Pinedale Field Office.

required about 4 acres per well and usually contained one or two wells. New multi-well sites approved in the RMP are permitted to have up to 60 wells per site. As of 2012, long-term disturbance per well was 0.44 acres and total disturbance per well was 1.26 acres, with incremental improvements in these numbers each year.[5]

- A few new wells have also been drilled horizontally, and a small number of horizontal wells may be drilled in the future. Horizontal wells can tap a larger area of gas-rich rocks, so fewer wells are required to extract the gas.

- Recycling of produced water began in 2006. Idle gas-gathering pipelines were repurposed to move produced water to central processing facilities and move processed water to new hydraulic fracturing sites. This system removed the need for hundreds of thousands of truck trips and reduced air pollution and wildlife disruption.[13] Several years later, water processing facilities were updated to clean some of the produced water to drinking water standards. This water is used for cement and other field operations that demand fresh water. Large amounts of treated water are also released into the local river – over one million gallons in the first year of operation.[14] Such water treatment and re-use practices are particularly important in semi-arid southwest Wyoming.

- More efficient drilling techniques have reduced drilling times from 45-50 days per well to 10-11 days per well, which reduces some environmental impacts such as air pollutant emissions from engines and truck traffic.

- The shift to year-round operations in designated development areas encourages a stable, less transient workforce, which benefits nearby communities.

- The Pinedale field is part of the Upper Green River Basin Ozone Nonattainment Designation Area designated by the U.S. Environmental Protection Agency (EPA) in 2012. In response to this designation, Wyoming expanded its efforts to reduce ozone levels in the area. State regulations included emission controls on storage tanks, pneumatic controllers, and drilling rig engines, as well as "green" well completions that capture the gases and fluids produced as a well is cleaned of fluids and debris after hydraulic fracturing.[15] These requirements helped reduce ozone levels in the Pinedale area, and in 2016 the EPA determined that the Green River Basin area had met the required ozone standards.[16]

- BLM rules require ongoing site reclamation. When active drilling operations end, well sites need less space and smaller access roads. Operators are required to promptly restore and revegetate unused areas under BLM oversight.

References & More Resources

For a complete listing of references, see the "References" section of the full publication, *Petroleum and the Environment*, or visit the online version at: www.americangeosciences.org/critical-issues/petroleum-environment

Wyoming State Geological Survey – Cultural Geology Guide: Pinedale Anticline. http://www.wsgs.wyo.gov/public-info/guide-pinedale

U.S. Bureau of Land Management (2008). Pinedale Resource Management Plan. https://eplanning.blm.gov/epl-front-office/eplanning/planAndProjectSite.do?methodName=dispatchToPatternPage&currentPageId=88620

U.S. Department of Energy, Office of Oil and Natural Gas – Footprint Reduction. https://www.energy.gov/sites/prod/files/2016/07/f33/Footprint%20Reduction.pdf

Petroleum and the Environment
Part 11

Heavy Oil
Abundant but hard to work with, heavy oil has some specific environmental impacts

Introduction

Naturally occurring crude oil comes in many forms. The most familiar to many people is light crude oil, which is less dense than water and flows easily at room temperature. Heavy oil and bitumen are forms of crude oil that are more viscous (thicker) and dense. The largest crude oil deposits in the world are heavy oil, extra-heavy oil, and bitumen oil sands (also called tar sands) in Venezuela and Canada. The U.S. also has heavy oil and oil sands, mostly in California, Alaska, and Utah. Globally, almost 1.1 trillion barrels of heavy oil, extra-heavy oil, and natural bitumen may be technically recoverable, compared to 950 billion barrels of light crude oil.[1]

Vast heavy oil resources pose an environmental conundrum: they are major energy resources and important to their host countries' economies, but they require more energy and water to produce and refine than lighter oils. They also contain sulfur and a range of polluting or toxic contaminants, including heavy metals, which must be removed and disposed of, further increasing costs and environmental impacts.[1,2]

Production Techniques

Because heavy oils are very viscous, they are difficult to extract from rocks. Different techniques are used depending on the type of oil and the properties and depth of the rocks:

Oil sand from Athabasca, Canada. The oil in these sands is so thick (viscous) that special processing is required to separate it from the sand. Image credit: Wikimedia Commons user Int23.[3]

- **Open-pit mining** – used for oil sands that are very close to the Earth's surface (typically less than 250 feet deep). The oil sands are mined in bulk, crushed, and transported to processing facilities that separate the oil from the sand using hot water and/or solvents. The ultra-thick oil (bitumen) is then refined or diluted with light oil for pipeline transport.[4] Open-pit mining is used for about 20% of Canadian oil sand production.[4] The Uinta basin in Utah also contains large, shallow oil sand deposits, but many efforts to produce oil from these sands have failed commercially.[5]

- **Injection of water, steam, and/or solvents** – used where heavy oil is deep below the surface, or where surface mining is not viable for environmental or commercial reasons. **Waterflooding** – the injection of water through one well to push oil towards another well where it is extracted – has been used to produce over 100 million barrels of heavy oil in Alaska since the early 1990s.[5] **Steam flooding** works in the same way, but the steam's heat softens the oil, allowing the process to be used for more viscous oils than waterflooding. This method is used in central California[6] and parts of Alberta. A special steam injection method called **steam-assisted gravity drainage (SAGD)** is used for 80% of Canadian oil sand production. SAGD involves the injection of steam into a horizontal well at the top of the oil sands. The heated and thinned oil then drains down into another horizontal well at the base of the oil sands, which then pumps the oil to the surface.[4] Any of these processes may be enhanced by adding solvents to the water.

- **Cold heavy oil production with sand (CHOPS)**[7] – used for mushy heavy and extra-heavy oil sands that can be extracted in their entirety through a well using intensive pumping. The oil, water, and sand are then separated at the surface. This technique has been tested in oilfields in Alaska's North Slope but not yet commercially developed due to low oil prices.[5]

Petroleum and the Environment
Part 11: Heavy Oil

Environmental Impacts Specific to Heavy Oil

Energy – heavy oils require much more energy to produce and refine than light crude oil. This leads to higher overall greenhouse gas emissions per barrel of oil produced, especially due to gas-fired steam generators and the energy-intensive processing required to lighten or break down heavy oil into forms that can be transported and used. Total "lifecycle emissions" from production, refining, transportation, and use for light vs. various heavy oils are:[8]

- Typical light West Texas oil - 480 kg CO_2 per barrel
- Canadian oil sands bitumen produced by SAGD, and Venezuelan extra-heavy oil, both diluted with lighter oil for ease of transport – 600 kg CO_2 per barrel
- Heavy oil produced by steam injection in California's Midway Sunset field - 725 kg CO_2 per barrel
- Canadian oil sands produced by open-pit mining and upgraded to a light synthetic crude oil ("syncrude") before transporting – 729 to 736 kg CO_2 per barrel

Open pits – open-pit mining of oil sands poses some specific environmental challenges that are less common elsewhere in the oil industry:

- Large volumes of tailings (residual clay, bitumen, and other chemicals) are stored in open surface ponds, presenting a potential risk to wildlife[9] and groundwater.[10,11]
- Tailings ponds, piles, and exposed heavy oil in the open mine, along with the heavy industrial activity common to all mining operations, are a major source of air pollution,[12] and dust from the mines can contaminate nearby surface waterbodies.[9]
- Open-pit mining of oil sands disturbs more of the land surface than oil wells. This impact is temporary if the mine land is fully reclaimed after the oil sands are extracted (as is currently required by the Government of Alberta, Canada), but has the effect of fragmenting or destroying habitats.[13]

Consistency of Heavy Oils

Heavy oil – like molasses
Extra-heavy oil – like peanut butter
Oil in oil sand – like window-sealing caulk or putty

Open-pit mining of oil sands in Alberta, Canada. The ponds in the photo are "tailings ponds", containing a mixture of water, fine sand, clay, and residual oil components after the sands have been processed to remove most of the oil. Image Credit: Dru Oja Jay, Dominion.[14]

U.S. Imports of Heavy Oil

The United States is the largest consumer of Canadian and Venezuelan heavy oil, extra-heavy oil, and bitumen. In 2017, the United States imported 2.7 million barrels of heavy oil per day from Canada[15] and 618,000 barrels per day from Venezuela.[16] Heavy oil imports from these two countries represented over 40% of U.S. crude oil imports in 2016.[16]

References & More Resources

For a complete listing of references, see the "References" section of the full publication, *Petroleum and the Environment*, or visit the online version at: www.americangeosciences.org/critical-issues/petroleum-environment

Gosselin, P. et al. (2010). Environmental and Health Impacts of Canada's Oil Sands Industry. The Royal Society of Canada Expert Health Panel, December 2010. https://rsc-src.ca/en/expert-panels/rsc-reports/environmental-and-health-impacts-canadas-oil-sands-industry

Natural Resources Canada – Environmental Challenges. https://www.nrcan.gc.ca/energy/oil-sands/5855

Carnegie Endowment for International Peace – Oil-Climate Index: Profiling Emissions in the Supply Chain. http://oci.carnegieendowment.org/#supply-chain

Petroleum and the Environment
Part 12

Oil and Gas in the U.S. Arctic
Managing resources in an oil- and gas-rich but harsh and fragile environment

Introduction

The Arctic hosts large oil and natural gas resources both onshore and offshore.[1] However, the harsh climate, extreme weather, remote locations, and limited infrastructure make exploration and production expensive and sometimes hazardous. In recent decades, decreased summer sea ice has resulted in increased shipping traffic and may encourage more offshore oil and gas exploration and production. Many concerns over the environmental impact of these activities are based on the Arctic's fragile, undisturbed ecosystems and the difficulty of monitoring and responding to spills due to remote locations, long, cold winters, and the lack of an Arctic deepwater port to handle emergency response vessels and equipment.[2]

In 2018, the intersection between oil, gas, and the Arctic environment is a topic of much current discussion due to recent federal decisions to expand federal oil and gas leasing in the coastal area of the Arctic National Wildlife Reserve (the "1002 area"), in a larger

Arctic Resources by the Numbers[1,33]

- The Arctic accounts for 6% of the Earth's surface but 10% of the conventional oil and gas discovered and produced to date.
- Undiscovered but recoverable "conventional" resources in the Arctic are estimated at 90 billion barrels of oil (16% of global total), 1,669 trillion cubic feet of gas (30%), and 44 billion barrels of natural gas liquids (38%). 84% of this is offshore.
- In 2016, Alaska's North Slope region produced about 5.5% of U.S. oil: 173 million barrels onshore and 6.2 million barrels offshore.
- North Slope peak oil production in 1988 was 720 million barrels, over 23% of all U.S. production.
- Alaskan oil production dropped over 75% from 1988 to 2015 as fields discovered in the 1960s were gradually depleted. New discoveries may change this trend.

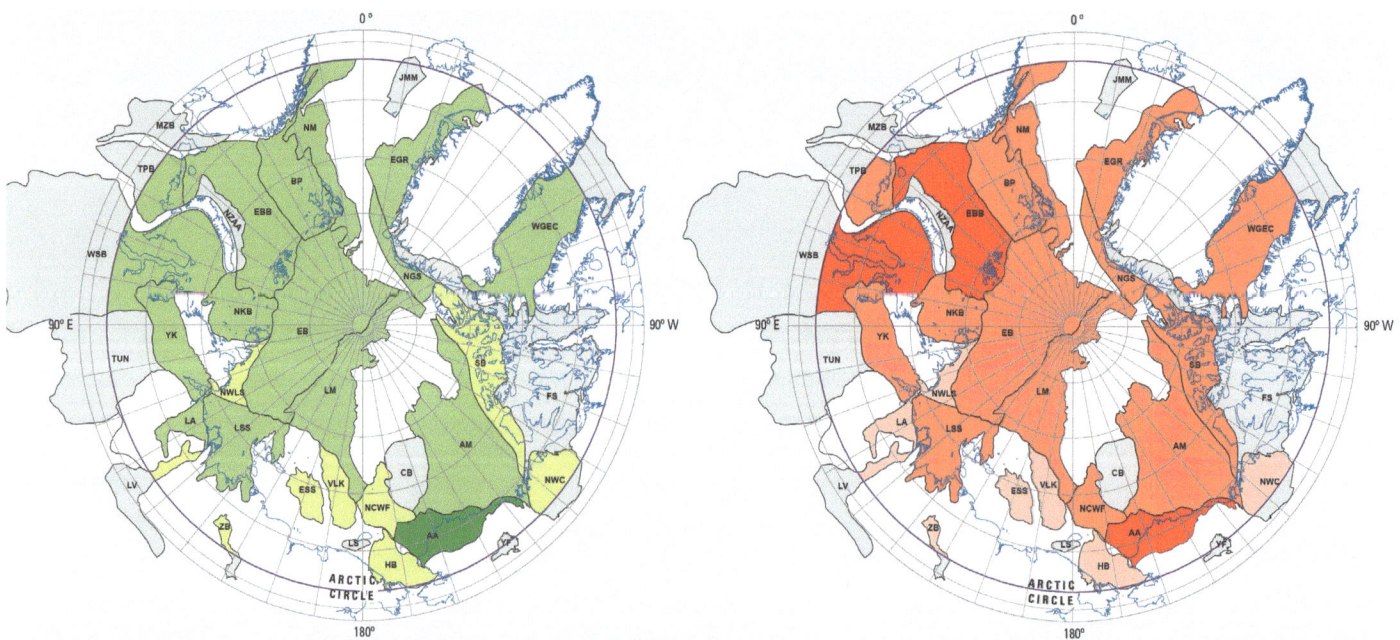

Estimated undiscovered oil (green, left) and gas (red, right) in the Arctic, according to the U.S. Geological Survey (2008). Greenland is in the top-right of each map, Alaska in the bottom-right ("AA" region covers Alaska's North Slope). Colors: undiscovered oil (greens) – dark = >10 billion barrels, medium = 1–10, light = <1; undiscovered gas (reds) – dark = >100 trillion cubic feet; medium = 6–100; light = <6. Image credit: U.S. Geological Survey[1]

Petroleum and the Environment
Part 12: Oil and Gas in the U.S. Arctic

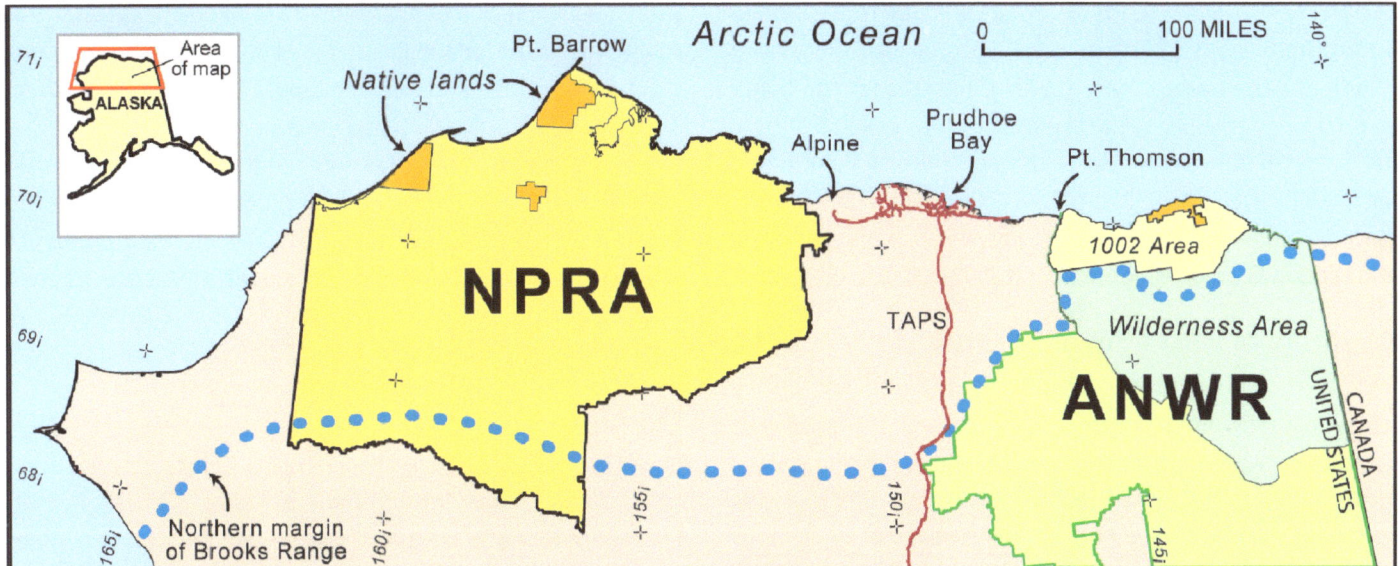

Oil and gas in the North Slope region of Alaska (north of the dotted blue line). Red lines are pipelines. Federal lands include the National Petroleum Reserve–Alaska (NPRA) and the Arctic National Wildlife Refuge (ANWR), which includes wilderness (darker green) and 1002 (light yellow) areas.[4] See text for more information about the features in this map. Image credit: U.S. Geological Survey[5]

area of the National Petroleum Reserve-Alaska (NPRA), and offshore.[3] Concerns over increased environmental risks in a region already experiencing rapid environmental and climatic changes are often weighed against the economic benefits of increased exploration and production. Attempting to balance these priorities, the National Environmental Policy Act (NEPA) requires that, before any exploratory studies are authorized, federal agencies must consider impacts on sociocultural, economic, and other natural resources in consultation with other government agencies and the public.

Land Ownership and U.S. Arctic Development

Although the federal government controls 60% of Alaskan land and all the ocean between the limits of state and international waters (i.e., from 3 to 200 nautical miles offshore), much of the oil and gas development to date has been on state and Native land around Prudhoe Bay in Alaska's North Slope.

National Petroleum Reserve-Alaska (NPRA):[4] The Reserve is a 23.6-million-acre tract of federal land set aside in 1923 to ensure future oil supplies for the U.S. Navy. Oil and gas resources in NPRA are significant but difficult to pinpoint because estimates change as exploration proceeds and additional data become available.[6] Leasing for oil and gas production in NPRA began in the 1980s, but development has been slow due to complex regulatory processes and the difficulty of operating in environmentally sensitive areas far from established infrastructure.

Arctic National Wildlife Refuge (ANWR): The Refuge was initially protected in 1960; in 1980, it was expanded and parts were designated as wilderness to preserve their unique wildlife, wilderness, and recreational value.[7] At the same time, Congress set aside an area on the North Coast (the "1002 area") to allow for future development of potentially large oil and gas resources.[8] The tax reform bill of 2017 (Public Law 115-97) opened the 1002 area to energy development.[9] This is the first time since 1980 that the 1002 area has been open to leasing or any activity (such as geophysical surveys) that could lead to producing oil or natural gas.[3]

State and Native Lands: 28% of Alaska is state-owned and 12% is Native land. As of early 2018, all onshore Alaskan North Slope oil production is on either state or Native lands, including from Native lands on the boundaries of NPRA. Production from federal lands in NPRA is expected to start in the near future.

Offshore: The state of Alaska regulates resources up to 3 nautical miles from the coast. From 3 to 200 nautical miles offshore, resources are regulated by the federal government.[10] One

Petroleum and the Environment
Part 12: Oil and Gas in the U.S. Arctic

notable offshore oilfield is the Northstar Oil Pool, which is located in both state and federal waters near Prudhoe Bay. Production in federal waters started here in 2001 from a man-made gravel island 6 miles offshore.[11] Future production is expected in both state and federal waters near existing onshore oil fields in the North Slope area. In 2017, the U.S. Department of the Interior started a multi-year process to develop a new five-year leasing plan that would allow large areas of offshore Alaska to be leased.[12]

Regulation of U.S. Arctic Drilling

Oil and gas development on federal land is regulated by the U.S. Bureau of Land Management. On state and Native lands, and in state waters, drilling and production of oil and gas and the underground disposal of oilfield waste are overseen by the Alaska Oil and Gas Conservation Commission.[13]

In federal waters, the U.S. Bureau of Ocean Energy Management (BOEM) manages leasing, including resource assessments to estimate potential lease value, and review and approval of drilling plans, including any necessary environmental assessments. The Bureau of Safety and Environmental Enforcement (BSEE) regulates all offshore drilling and production activities. The Arctic Drilling Rule released by BSEE in 2016[14] requires rigorous safety controls beyond those required in other offshore areas. These include having equipment on hand to cap an out-of-control well and capture any leaking oil, and having access to a separate rig that can drill a relief well and plug a compromised well permanently before seasonal ice encroaches on the drill site or within 45 days, whichever is sooner. Although improvements have been made, the safety and environmental impacts of offshore oil and gas development in the Arctic remain highly controversial.[15,16]

Reducing Surface Impacts on Alaska's North Slope

Oil production on the North Slope of Alaska began in 1977. The developed area is focused on a narrow coastal strip running about 100 miles east to west in the vicinity of Prudhoe Bay (see map). In the 1970s and 80s, access roads, well sites, oil and gas processing facilities, and support facilities were constructed using gravel. By the early 2000s, roughly 9,200 acres were under gravel.[17] More recent exploration reduces the surface impact by using ice roads and ice drilling sites that are constructed each winter, and small gravel production sites with multiple wells per site. For example, the Alpine field, which was discovered in 1994 to the west of the older oilfields,[18] uses about 100 acres of gravel drill pads, facilities, and roads to tap a 25,000-acre oil reservoir – as of early 2018, the field had produced 465 million barrels of oil.[19] In 2018, Greater Mooses Tooth-1, a field located in NPRA, west of Alpine, will begin producing oil from a single 12-acre gravel drill pad designed to support 33 wells.[20] Eight miles of gravel road and parallel above-ground pipeline, including two bridges, will connect the pad to existing Alpine field facilities, for a total gravel footprint of 73 acres.

Oil Pipelines

North Slope oil is collected by a network of local pipelines and then sent 800 miles through the Trans-Alaska Pipeline System (TAPS) to the south coast of Alaska. Except for small quantities refined in Alaska, most of the oil is loaded onto tankers and shipped to refineries on the West coast or occasionally in Hawaii. In areas where the soil is either permanently frozen (permafrost) or never freezes, the Trans-Alaska Pipeline is buried; in areas where the ground freezes and thaws with the seasons, the pipeline is generally elevated above ground.[21] Where highways, animal crossings, or unstable hillslopes required pipeline burial in unstable permafrost, insulation or refrigeration is used to keep the ground cold.[22] Safety features include systems that monitor variations in pipeline flow and pressure to alert response teams

The Trans-Alaska oil pipeline is mounted on sliders where it crosses the Denali fault. During a large (magnitude 7.9) earthquake in 2002, this system allowed the pipeline to move without breaking. Image credit: Tim Dawson, U.S. Geological Survey[25]

Petroleum and the Environment
Part 12: Oil and Gas in the U.S. Arctic

to the location of probable leaks;[22] tracks that allow the pipeline to move without breaking during earthquakes (see photo);[23] and heat transfer pipes that move heat from the buried pipe to the air, helping keep permafrost cold and stable.[22]

Oil Spills

A 2013 BOEM analysis of oil spills in the North Slope area between 1971 and 2011 identified 1,577 spills larger than 42 gallons (one barrel), 10 spills larger than 21,000 gallons, and two spills larger than 42,000 gallons.[25] In 2017, the Alaska Division of Spill Prevention and Response recorded 18 crude oil spills on the North Slope, releasing a total of 1,010 gallons of oil.[26] Spills of other operational fluids are more common: in 2017 there were 147 recorded spills of diesel, engine lube oil, gasoline, hydraulic oil, and produced water on the North Slope, totaling 43,000 gallons.

The largest North Slope oil spill occurred in 2006, when a pipeline leaked 267,000 gallons of crude oil onto the tundra in the Prudhoe Bay field. The leak was caused by the operator (BP)'s failure to prevent internal corrosion in a 29-year-old pipeline that had not been properly maintained and inspected. BP was fined $25 million and required to implement a system-wide pipeline integrity management program.[27] Cleanup operations included removing oil from snow and vegetation without disturbing the tundra and underlying soils, preventing the spread of leaked oil into nearby lakes, and restoring the site of the spill.[28]

The U.S. Bureau of Safety and Environmental Enforcement supports research into offshore spill response in the presence of sea ice, cold temperatures, and the hazardous conditions of the Arctic. Studies on improved spill cleanup include mechanical cleanup, in-situ burning, and chemical treating agents. Research also investigates improved oil detection methods for locating spills.[29]

In 1989, the Exxon Valdez oil tanker was carrying North Slope oil when it ran aground and ruptured, spilling 11 million gallons of crude oil into Prince William Sound in southern Alaska. This spill was the largest oil spill in the United States until the 2010 Deepwater Horizon disaster, which was almost 20 times larger. Shipping regulations developed in response to the spill, and supplemented in 2009, primarily focus on Alaskan oil exports. Regulations now require double-hulled tankers, tugboat escorts with spill response capability, U.S. Coast Guard electronic tracking throughout Prince William Sound and 60 miles into the Gulf of Alaska, and local availability of response and recovery equipment and trained personnel.[30] The National Oceanic and Atmospheric Administration continues to monitor the ongoing wildlife recovery from the Exxon Valdez spill.[31]

Who are the Arctic Oil & Gas Producers?

As of 2018, three nations produce oil and gas north of the Arctic Circle: the U.S. (Alaska), Russia, and Norway. Canada, Greenland, Iceland, Sweden, and Finland currently have no Arctic production. Production may expand to other Arctic nations in the future. Canada, Finland, Iceland, the Kingdom of Denmark (Greenland), Norway, Russia, and the United States exchange information, best practices, and regulatory experiences through the Arctic Offshore Regulators Forum.[32]

References & More Resources

For a complete listing of references, see the "References" section of the full publication, *Petroleum and the Environment*, or visit the online version at: www.americangeosciences.org/critical-issues/petroleum-environment

National Academies Ocean Studies Board (2014). Responding to Spills in the U.S. Arctic Marine Environment. http://dels.nas.edu/Report/Responding-Spills/18625

Arctic Council. Intergovernmental forum for sustainable development and environmental protection in the Arctic: http://www.arctic-council.org/index.php/en/

National Petroleum Council (2015). Arctic Potential: Realizing the Promise of U.S. Arctic Oil and Gas Resources. http://www.npcarcticpotentialreport.org

Pew Charitable Trusts – Arctic Standards http://www.pewtrusts.org/~/media/assets/2013/09/23/arcticstandardsfinal.pdf

Petroleum and the Environment
Part 13

Offshore Oil and Gas
Technological and environmental challenges in increasingly deep water

Introduction

Many of the world's oil and gas resources lie beneath the oceans. Advances in exploration, drilling, and production technologies allow production in water more than 10,000 feet deep and more than 100 miles offshore. Major spills are rare but damage sensitive ocean and coastal environments, affect local economies, and are difficult and expensive to clean up. Federal regulations and industry standards have advanced to improve the safety and reduce the environmental impacts of offshore oil and gas production, particularly since the 2010 *Deepwater Horizon* disaster and oil spill. As drilling and production become possible under increasingly extreme physical conditions, the issues surrounding environmental protection and safety in offshore oil and gas continue to evolve.

Seaward Progress of Oil and Gas Exploration

Over the last 120 years, offshore drilling has advanced seaward from drilling rigs mounted on shoreline piers, to rigid platforms mounted on the seafloor, to floating and seafloor systems (see figure) in water depths up to 10,000 feet. The complexity and high cost of drilling in deep water – several hundreds of million dollars per well – can be justified by the high productivity of the oil fields if the oil price is sufficiently high to make them profitable.[1] Oil and gas exploration has also moved offshore in the Arctic, an area with shallow water depths but severe weather hazards. The challenges relating to Arctic offshore drilling differ substantially from those in other offshore regions – see "Oil and Gas in the U.S. Arctic" in this series for more on this topic.

Economic Constraints

Deepwater oilfield development may take ten years from the first exploratory well to the first barrel of oil sold, and pre-production costs, including development wells and specially designed production facilities, may be several billion dollars.[1] Safe and successful offshore drilling and production requires extensive seismic imaging and geologic analysis, engineering design and planning, construction of highly specialized equipment, and compliance with federal or state environmental regulations, all

Offshore Oil and Gas Production in Brief

- U.S. offshore oil production in 2017: 602 million barrels (18% of U.S. total)[2]
- U.S. offshore natural gas production in 2016: 1.7 trillion cubic feet (5.2% of U.S. total)[3]
- Federal revenues from offshore oil and gas (2017): $3.8 billion[4]
- Globally, offshore oil production in 2015 was roughly 10 billion barrels, about 29% of total global production.[5]
- Largest offshore producers: Saudi Arabia, Brazil, Mexico, Norway, U.S.[5]
- Largest producers in deepwater (>125 m, roughly 400 ft) and ultra-deepwater (>1500 m, roughly 5000 ft): Brazil, Angola, Norway, U.S.[6]

U.S. offshore oil and gas production is focused in the central and western Gulf of Mexico, with some production off the coast of southern California and in the Cook Inlet and Beaufort Sea of Alaska.

Globally, offshore leaks and spills account for a small amount of the total oil that gets into ocean waters. Natural oil seeps on the seafloor contribute up to half of the oil in the ocean, although these are distributed widely and so do not have the same local effects as a large spill. Other sources include boat engines, discharged ballast water from tankers, contaminated river water and wastewater drainage on land, and oil drips and exhaust from automobiles.[7]

Major spills are rare but can cause extensive damage to sensitive ocean ecosystems due to the large amount of oil leaked in a small area.

of which take a lot of time and money. Companies therefore make decisions about offshore drilling and development based on expected future (rather than current) oil prices. Offshore oil

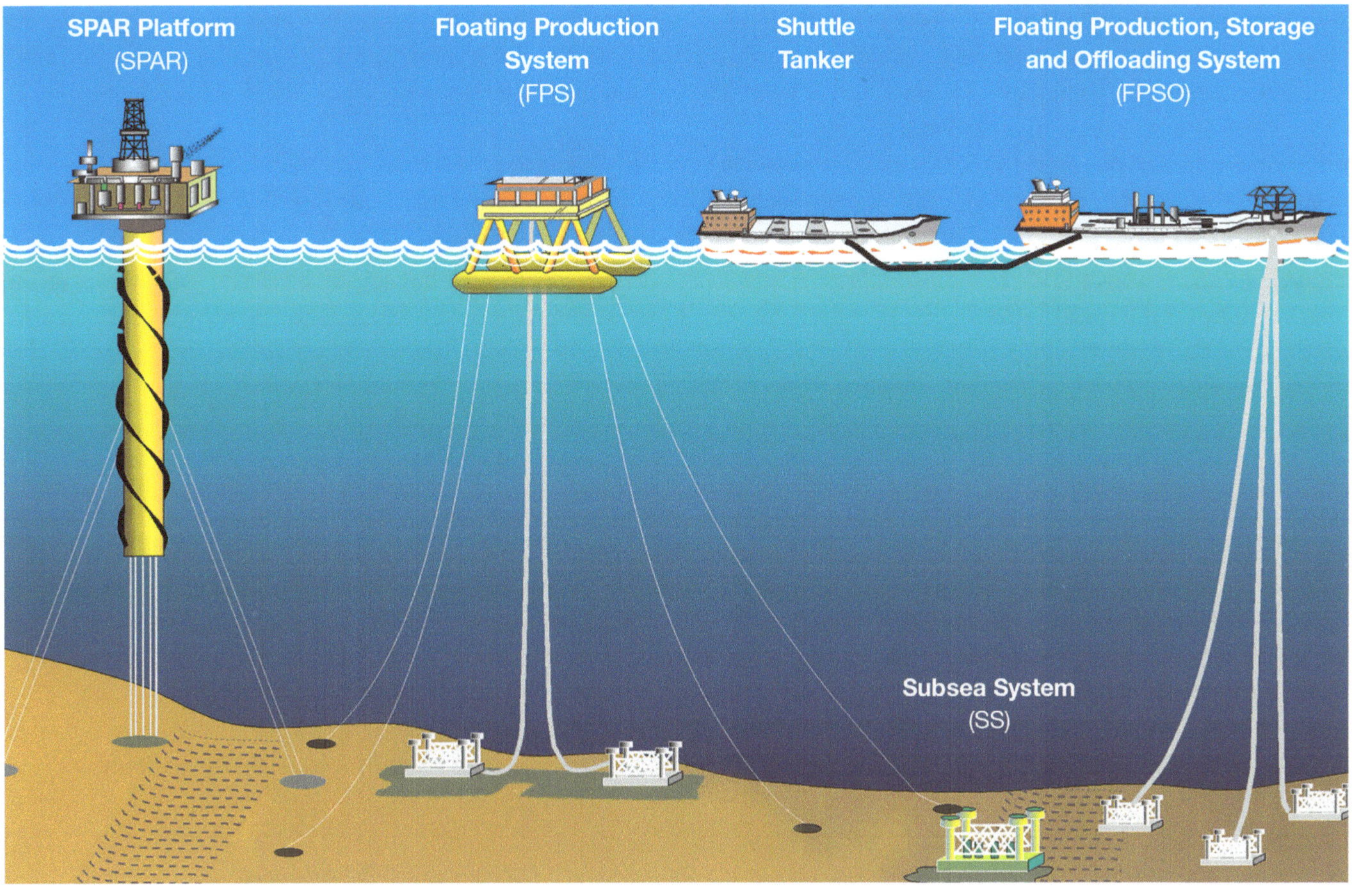

Petroleum and the Environment
Part 13: Offshore Oil and Gas

A wide variety of technologies are used to drill offshore in increasingly deep water. Image credit: Bureau of Ocean Energy Management.[2]

production therefore does not respond to oil prices in the same way as onshore production. For example, while oil prices fell and onshore U.S. oil production plateaued in 2014-2015,[9] production on the Gulf of Mexico Outer Continental Shelf (OCS) increased by almost 25% from November 2014 to December 2016.[10]

Technological Advances in Offshore Drilling

Drilling and producing technologies for progressively more complex operating environments are often developed by collaborations between industry, service companies, academia, and research institutions sponsored by the federal government[11] or industry.[12] Some of the deepest Gulf of Mexico wells are in water more than a mile deep, with some wells extending more than 20,000 feet below the seafloor.[13] Companies are planning to explore in deeper and more hazardous regions in the U.S. and around the globe, which will depend on continuing technological advances. Frontiers in this area include:

- **Seismic imaging beneath salt layers** – in the Gulf of Mexico and offshore Brazil, some oil reservoirs are found in complexly folded and faulted rock formations beneath thick layers of salt. Salt layers reflect seismic waves, making it difficult to image underlying rock layers with those waves. New analytical techniques are continuously being developed to improve imaging beneath salt layers, using supercomputers to process huge quantities of seismic data.[14] Salt also dissolves in drilling fluids, and at high temperatures and pressures the salt itself can flow, squeezing and damaging the wellbore and drilling equipment, so drilling through salt requires advanced well stabilization and drilling techniques.[15]

- **High-temperature, high-pressure materials and electronics** – current technologies allow for drilling at temperatures up to 350°F, but future deep wells may require operating temperatures as high as 500°F: good for baking pizza, but

Petroleum and the Environment
Part 13: Offshore Oil and Gas

not so friendly for sophisticated electronics or drilling mud. Wells can currently operate under very high pressures, up to 15,000 pounds per square inch (psi), but future ultradeep wells will be expected to withstand as much as 30,000 psi (2000 times atmospheric pressure).[16] Very high temperatures require special materials for drilling, but the major constraint on operating in these extreme conditions may be the fragility of the electronics that guide directional drilling equipment inside the well.[17]

- **Installation and monitoring** – deepwater production is shifting from floating platforms to seafloor wellheads connected to seafloor pipelines. These systems rely on improved autonomous installations and remote monitoring equipment, including unmanned underwater vehicles.[18]

- **Preventing well blowouts** – especially since the *Deepwater Horizon* disaster, an ongoing concern is ensuring that well blowout preventers (BOP) are reliable, and that well designs and operations are safer to reduce the need for a BOP (see "U.S. Regulation of Oil and Gas Operations" in this series).[13]

- **Hurricanes** regularly cross oil-rich parts of the Gulf of Mexico and are expected to increase in intensity in the future.[19] Improvements to equipment reliability and performance, and remote underwater systems, will be key to improving the resilience of offshore oil and gas infrastructure.

Transocean Development Driller drilling a relief well in the Gulf of Mexico to relieve the pressure on the leaking Macondo well, 2010. This rig can drill to depths of 37,500 feet in as much as 7,500 feet of water. Image credit: U.S. Coast Guard.[20]

- Improved systems to prevent the drilling mud that travels between the drilling rig and the seafloor wellhead from putting too much pressure on the well itself.[13]

Oil Spills

Major offshore oil spills are rare but can cause great harm to coastal and marine wildlife and the people who depend on marine and coastal resources. Developments in offshore environmental policy and activity over the last 50 years have been punctuated by a few major events:

- In 1969, an offshore well blowout near Santa Barbara, California, deposited thick layers of oil along 35 miles of coastline and killed thousands of birds and marine animals. The spill paved the way for major environmental laws and was one of the events that led to the establishment of the U.S. Environmental Protection Agency in 1970.[21]

- Three shipping oil spills in 1989, including the Exxon Valdez accident in Alaska, prompted developments in environmental-protection regulations, including the requirement for all new tankers to have double hulls.[22]

- In April 2010, the Macondo well blowout and oil spill occurred 50 miles off the coast of Louisiana in approximately 5,000 feet of water. Eleven crew on the *Deepwater Horizon* drillship were killed and approximately 4.9 million barrels of oil were spilled (205 million gallons, equivalent to 320 Olympic-size swimming pools or 3 days of all Gulf of Mexico oil production). In response to the spill, industry and state & federal agencies launched massive cleanup operations; a National Academies assessment of the spill recommended changes for the industry and regulators;[23] and federal regulatory agencies were reorganized.[24] Regulators have continued to issue revised requirements for equipment and procedures from 2012 to the present day. In 2016, a $20.8 billion environmental damage settlement was reached between the United States, the five Gulf states, and the well operator (BP).[25] This settlement provided $1.86 billion for ecological and economic recovery; $1.6 billion for region-wide restoration; over $130 million for research and technology development by the National Oceanic and Atmospheric Administration to support Gulf ecosystems,

Petroleum and the Environment
Part 13: Offshore Oil and Gas

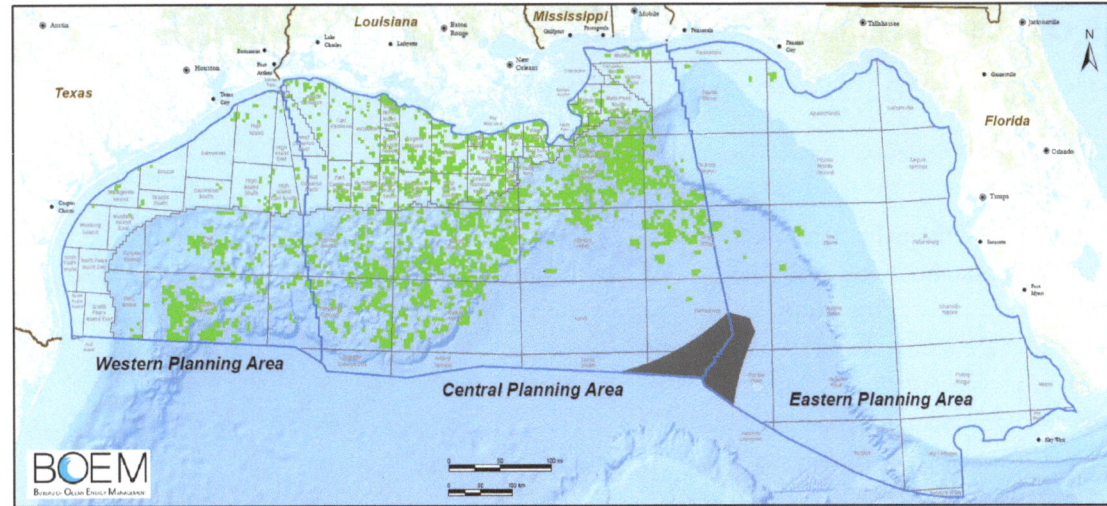

Planning areas (blue), region blocks (grey), and active leases (green) in the Gulf of Mexico as of February 1, 2018, as administered by the U.S. Bureau of Ocean Energy Management. There are several thousand leases in the Gulf of Mexico, covering almost 15 million acres. Image credit: U.S. Bureau of Ocean Energy Management.[26]

recreation, and fishing; and $133 million to establish "Centers of Excellence" for Gulf science and technology. BP also provided $500 million in 2010 for the Gulf of Mexico Research Initiative, a 10-year effort to improve understanding of the environmental stresses and public health implications of spill events.[27] The Gulf of Mexico Research Institute and others continue to assess the fate of the spilled oil and impacted land and wildlife.[28]

Some large spills have occurred in the Gulf of Mexico since the Deepwater Horizon disaster. An underwater pipeline near the Delta House production facility, 40 miles southeast of Venice, Louisiana, is estimated to have spilled as much as 672,000 gallons in October 2017.[29,30] Most of the oil dispersed into the surrounding ocean rather than reaching the shore. In 2004, the Taylor Energy Mississippi Canyon 20A platform, with 25 connected wells, was seriously damaged by Hurricane Ivan. Since then, the company, the U.S. Coast Guard, and several federal agencies have worked to remove the platform, decommission the oil pipeline, and decommission 9 of the 25 connected wells (as of early 2015) using funds provided by Taylor Energy. The remaining wells continue to leak oil into the Gulf of Mexico: in 2014-2015, it was estimated that the leak rate varied from 42 to 2,329 gallons per day.[31]

Regulation

Individual states control waters from the coast outward to 3 to 9 nautical miles, depending on the state. Federal regulation covers the Outer Continental Shelf (OCS) – the area beyond state waters out to 200 nautical miles offshore, or to the border with another country's exclusive economic zone (e.g., in the Gulf of Mexico). The U.S. Bureau of Ocean Energy Management, established in 2011, issues OCS leases for oil, gas, and wind energy development. The U.S. Bureau of Safety and Environmental Enforcement (BSEE), also established in 2011, regulates energy activities on the OCS. Other federal agencies contribute biologic, geologic, environmental, and security expertise and regulatory authority.

BSEE issued its Well Control Rule in 2016 to improve the effectiveness of the equipment that prevents well blowouts.[32] Other 2016 rules included updated regulations for production facilities and equipment, and for Arctic drilling. All rules are developed with public input, including public comment sessions. As of 2018, these regulations are being reviewed and revised by the Administration.

In 2016, President Obama banned new offshore oil and gas activities in parts of the Atlantic and Alaska. In 2017, the Administration started work on a revised offshore leasing plan that may expand areas open for oil and gas development.[33]

References & More Resources

For a complete listing of references, see the "References" section of the full publication, *Petroleum and the Environment*, or visit the online version at: www.americangeosciences.org/critical-issues/petroleum-environment

Petroleum and the Environment
Part 14

Spills in Oil and Natural Gas Fields
Spill types, numbers, sizes, effects, and mitigation/cleanup efforts

Introduction
Oilfield spills can harm wildlife and pose a risk to human health if they reach fresh water sources or contaminate soil or air. The enormous size of the oil and gas industry and the huge volumes of oil and produced water that are handled, stored, and transported result in thousands of spills every year.[1,2,3] But not all spills are created equal: the size, location, and type of spill, and how quickly the spill can be cleaned up, all influence the overall environmental impact.

Spills occur in two main settings: at or near the well site, or in transit between the oilfield, refineries, and consumers. Spills in the oilfield are usually smaller[4] and easier to clean up than those related to bulk transportation: drill sites are purpose-built "pads" made of gravel and other materials designed to deter spills from reaching soil or groundwater; additional containment measures are used around liquid storage tanks or pits to help contain spills; and equipment and personnel are commonly on hand to address spills quickly.

Other parts of this series provide more information on spills related to transportation and offshore oil and gas production. Spill cleanup in the ocean is particularly complex and is not addressed in detail in this series. Instead, we refer the interested reader to the excellent introductory resources provided by the National Oceanic and Atmospheric Administration – see "References & More Resources" at the end of this section for more information.[5] Spill cleanup on land is described in the following pages, as similar methods are used both on and off the oilfield.

Number and Size of Spills
Spill data for different states are difficult to compare because each state has different reporting requirements based on the volume, type of material, location, and spread of the spill. A study of over 31,000 horizontal, hydraulically fractured wells in Colorado, New Mexico, North Dakota, and Pennsylvania from 2005-2014 found that, on average, there were 55 reported oilfield spills per 1000 wells per year.[2] Most spills are small: the

Spill from a leaking oil well in the Salt Wash oilfield, Utah, 2014. Image credit: U.S. Bureau of Land Management.[6]

median reported spill size was 120 gallons in Pennsylvania, 210 gallons in North Dakota, 798 gallons in Colorado, and 1,302 gallons in New Mexico (note again that state numbers should not be directly compared because of different reporting requirements in each state). Roughly half of all spills documented in this study came from storage tanks, storage pits, or the pipes that transport used drilling fluid around and away from a drill rig. Most spills involved produced water (the naturally occurring water – sometimes very salty – that is present in oil and gas reservoirs and so is extracted along with the oil/gas); in some cases this can include flowback of hydraulic fracturing fluid. The same study found that 75–94% of spills occurred during the first three years of a well's life. This is the time during which a well is drilled, hydraulically fractured, and has the highest production rate. These factors result in larger volumes of fluids being stored, handled, and transported, all of which increase the risk of a spill.

A U.S. Environmental Protection Agency (EPA) assessment of spill data from January 2006 to April 2012 studied 36,000 spills from state and industry data sources in an attempt to identify spills specifically related to the hydraulic fracturing process.

Petroleum and the Environment
Part 14: Spills in Oil and Natural Gas Fields

Of the 36,000 spills studied, roughly 24,000 were unrelated to hydraulic fracturing, 12,000 had insufficient data to make this determination, and 457 spills were documented as directly related to hydraulic fracturing.[7] Of these 457 spills, 40% were smaller than 500 gallons and 5% were larger than 10,000 gallons; half consisted of flowback and produced water; and almost half were from storage tanks or pits. The most common cause of spills was human error.

What is Being Spilled?

Many different fluids are handled in the oilfield. The most commonly spilled fluid is produced water (sometimes including flowback of hydraulic fracturing fluid). Other commonly spilled fluids include crude oil, fluids to be used for hydraulic fracturing, and drilling waste (slurries of rock chips and drilling mud produced during drilling).[8] For more information on the composition and handling of produced water, see "Using Produced Water" in this series. For detailed information on the composition of hydraulic fracturing fluids, the FracFocus database provides a registry of fluid compositions used in over 120,000 wells across the United States since 2011.[9]

Environmental Impacts of Spills

The environmental impact of a spill depends strongly on the size, location, type of fluid, and spread of the spill, including whether or not it contaminates ground- or surface water, which allows it to spread further and makes cleanup more difficult:

- Spilled oil or refined fuel can coat plants, soils, microbes, and animals. Oil prevents plant growth and hinders the movement of water, oxygen, and nutrients through soils. Some components of oils and liquid fuels are toxic to plants, animals, and humans.[10,11]
- Some light oils and refined fuels such as gasoline or diesel may evaporate, releasing toxic fumes that may degrade air quality or pose a fire hazard.
- Highly saline produced water (up to 15 times saltier than seawater) can kill vegetation and prevent plants from growing in contaminated soil.

Spill Prevention and Mitigation

States regulate oil and gas exploration and production operations and specify spill reporting and cleanup requirements. In addition, the EPA oversees the reporting and cleanup of spills

Storage tanks for produced water from natural gas drilling in the Marcellus Shale of Western Pennsylvania are surrounded by spill-containment structures. Storage tanks are one of the main sources of oil- and gas-field spills. Image credit: Doug Duncan, U.S. Geological Survey.[12]

that impact inland waters of the United States. The U.S. Coast Guard is the lead response agency for coastal waters and ports.

Some spills are caused by technical failures, but the most common cause of oilfield spills is human error. Safe and effective operating procedures are therefore an important part of spill prevention. All operators at sites that could potentially discharge oil into or near navigable waters are required by the EPA (or state regulators delegated by the EPA) to implement Spill Prevention, Control, and Countermeasure (SPCC) plans.[13] Equipment monitoring and the development of newer, safer equipment aim to reduce the number of spills due to technical and mechanical failures.

In areas prone to spills, such as wastewater storage tanks, containment methods such as berms and plastic sheeting are used to prevent spills from entering the surrounding environment. Similar measures can be set up quickly around a spill to reduce contamination and speed cleanup.

Oil Spill Cleanup

The first step in oilfield spills is typically to capture spilled fluid using booms, vacuum tanks, or absorbent material, and then to remove any contaminated rocks, plants, and soil.[14] If this is done quickly, spilled material can be prevented from

Petroleum and the Environment
Part 14: Spills in Oil and Natural Gas Fields

A contractor works to remove oil pools left after a spill from a refinery in Coffeyville, Kansas, contaminated floodwaters of the Verdigris River in 2007. Image Credit: Leif Skoogfors/U.S. Federal Emergency Management Agency.[15]

spreading further into soil or entering ground- or surface water. Of the 457 spills studied by the EPA from 2006 to 2012, 64% reached the soil, 7% entered surface water, and one spill entered groundwater.[14]

Oil and refined fuels are degraded by air, sunlight, and bacteria, so for these kinds of spills, fertilizers may be added to spill areas to speed up bacterial activity and more rapidly remediate the spill.[16] If an oil spill occurs in a contained area, the oil may be burned off, or the area may be flooded with water so that the oil can float to the water surface for easier removal. Larger animals such as birds, fish, and mammals can be individually collected, cleaned, and treated to prevent toxins from being absorbed through their skin, but this is expensive and has a low success rate.[17]

In 1979, 420,000 gallons of crude oil spilled from a pipeline in Bemidji, Minnesota. The area has since become the National Crude Oil Spill Research Site, managed by the U.S. Geological Survey's Toxic Substance Hydrology Program to investigate the effects of a land-based oil spill. Research conducted at this site is used to develop new remediation techniques.[18]

More detailed information about spill cleanup techniques used by the petroleum industry and their advantages and disadvantages can be found in the American Petroleum Institute's Oil Spill Prevention and Response Cleanup Toolkit - see "References & More Resources" for more information.[19]

Brine Spill Cleanup

State records do not give details about the composition of spilled water.[9] However, the U.S. Geological Survey database of produced waters shows that the majority are saline.[20] Saline water spills have long been known to kill existing vegetation and prevent revegetation, which can lead to deep soil erosion.[21] Produced water may also contain other toxic components, such as barium, and elevated concentrations of naturally occurring radioactive elements, such as radium.

In January 2015, three million gallons of produced water leaked from a shallow underground pipeline in the Williston Basin, North Dakota.[22] The produced water in the pipeline was almost nine times saltier than seawater and contained residual oil. The leak occurred near Blacktail Creek, posing a risk to aquatic life at the site and further downstream. Remediation was conducted by

Cleanup crew at the 2015 Blacktail Creek spill, Marmon, North Dakota. This spill was caused by a leaking pipeline that was transporting produced water in the oilfield. Image credit: U.S. Environmental Protection Agency.[24]

Petroleum and the Environment
Part 14: Spills in Oil and Natural Gas Fields

the pipeline operator under the oversight of the North Dakota Department of Health and included:[23]

- Recovering small quantities of oil from surface water
- Removing up to a foot of soil from several acres around the site
- Impounding Blacktail Creek to allow for the contained collection of contaminated water
- Pumping out potentially contaminated groundwater near Blacktail Creek for several miles downstream

Samples collected four months after the spill showed only slightly elevated salinity levels downstream from the spill. These levels were above background levels but well below the EPA's drinking water standards. However, some toxic elements (such as barium and radium) stuck to soil particles, reducing the immediate spread of those elements but presenting a longer-term risk if those elements are later remobilized by water passing through the soil. Overall, some minor harm to the creek ecosystem was observed, but fish survival rates appeared to be largely unaffected by the spill, suggesting that the remediation efforts were generally effective.

References & More Resources

For a complete listing of references, see the "References" section of the full publication, *Petroleum and the Environment*, or visit the online version at: www.americangeosciences.org/critical-issues/petroleum-environment

National Oceanic and Atmospheric Administration – How Do Oil Spills out at Sea Typically Get Cleaned Up? https://response.restoration.noaa.gov/about/media/how-do-oil-spills-out-sea-typically-get-cleaned.html

API Energy – Oil Spill Prevention and Response: Toolkit. http://www.oilspillprevention.org/oil-spill-cleanup/oil-spill-cleanup-toolkit

Information on research into the composition and risks of oilfield spills is available from the U.S. Geological Survey research program, *Understanding the Potential Risks to Water Resources and Environmental Health Associated with Wastes from Unconventional Oil and Gas Development*. https://toxics.usgs.gov/investigations/uog/more_uog/research_goals_and_approach.html

For even more detail on the sources, prevention, assessment, cleanup, and economic and legal context of spills, see: Testa, S.M. and Jacobs, J.J (2014). Oil Spills and Gas Leaks: Emergency Response, Prevention, and Cost Recovery. McGraw Hill Publishers, 578 p.

Petroleum and the Environment
Part 15

Transportation of Oil, Gas, and Refined Products
The methods, volumes, risks, and regulation of oil and gas transportation

Introduction

The U.S. has millions of miles of oil and gas pipelines, thousands of rail cars, vessels, and barges, and about 100,000 tanker trucks that move oil and gas from wells to processing facilities or refineries, and finally to consumers. The U.S. also imports and exports large volumes of oil, refined products, and natural gas by pipeline and ship. This vast transportation web includes regional and national storage sites for crude oil, heating oil, gasoline, and natural gas, which help with unexpected demand or delivery interruptions.

Across the U.S., natural gas is transported almost entirely by pipeline, and over 90% of crude oil and refined petroleum products are transported by pipeline at some point. If you have gas-powered heat in your home, there are almost certainly gas distribution lines running down your street and into your house.

If you live in a large city or a major oil-producing region, there are almost certainly oil and/or gas transmission pipelines in your area. Pipelines are all around us; like water pipes or electricity lines, they form a critical part of our national infrastructure and generally operate without most people noticing.

When pipelines are well constructed and maintained, spills and leaks are very rare, but no transportation system is 100% safe. Oil spills on land can have significant local impacts; spills in the ocean can have regional impacts; and gas leaks emit methane, which is a potent greenhouse gas and contributor to ozone pollution. Spills and leaks also waste resources (and therefore money), so there are environmental, public health, and economic incentives for maintaining a safe and efficient transportation system for our energy resources.

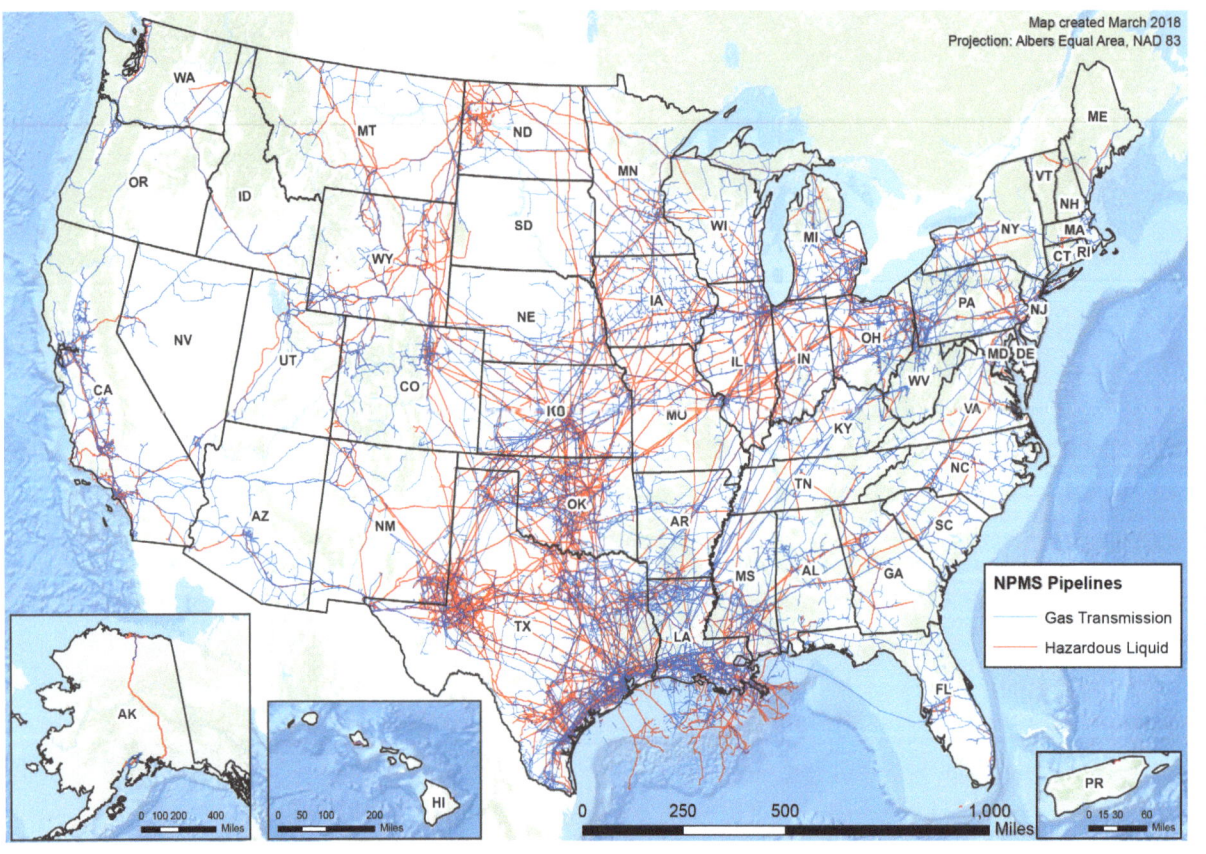

Gas transmission (blue) and hazardous liquid (red) pipelines in the United States as of March 2018. Image does not include gas distribution lines. Hazardous liquid pipelines are mostly those used for crude oil, refined products, and other petroleum liquids, but also include pipelines used to transport ammonia and carbon dioxide. Image credit: Pipeline and Hazardous Materials Safety Administration.[1]

AGI Critical Issues Program: www.americangeosciences.org/critical-issues
Supported by the AAPG Foundation. © 2018 American Geosciences Institute

Petroleum and the Environment, Part 15/24
Written by E. Allison and B. Mandler for AGI, 2018

Petroleum and the Environment
Part 15: Transportation of Oil, Gas, and Refined Products

Pipelines

The U.S. has over 200,000 miles of pipeline for crude oil, refined products, and natural gas liquids. There are over 300,000 miles of pipeline for gathering and transmitting natural gas, and 2.2 million miles for distributing gas to homes, businesses, and other industrial sites.[2]

In 2014, U.S. pipelines transported almost all of the natural gas produced and used in the United States, as well as over 16 billion barrels of crude oil and refined products[3] (this number is about twice the total U.S. consumption of crude oil and refined products because they are transported multiple times between well, refinery, and retail outlet[4]).

Except for above-ground shut-off valves, most pipelines are buried and marked by signs that warn against digging, identify the line location, and provide emergency contact information in case of a leak or spill.

Sign indicating a buried refined-product pipeline in Alexandria, Virginia. The Plantation Pipeline runs from Gulf Coast refineries to Northern Virginia; the pipeline shown in this picture is an extension of the Plantation Pipeline that delivers jet fuel to Washington National Airport. Image credit: Ben Mandler, American Geosciences Institute.

Rail

In 2017, roughly 140 million barrels of crude oil were transported by rail in the United States, a 64% decrease compared to 2014 volumes.[5] This decline was due to the opening of new crude oil pipelines and fewer shipments from the Midwest to the East Coast as coastal refineries imported more crude oil.[6] Crude oil imports by rail from Canada were about 48 million barrels in 2017.[7]

Trucks

In 2013, trucks carried about 1.2 billion tons (about 380 billion gallons or 9 billion barrels) of gasoline, diesel, and aviation fuel.[8] Many of the shipments were from refineries or bulk storage facilities to over 160,000 retail outlets.[9] Trucks are the most versatile form of transportation because they don't rely on the presence of pipelines, railways, or water. As a result, they are used for most short-distance transportation of oil and refined products. However, trucks are not particularly energy-efficient, requiring three times as much energy as a train (which is itself less efficient than a barge or pipeline) to move the same amount of material,[10] so trucks are less attractive for long-distance transportation.

Barges

Barges have traditionally moved small amounts of crude oil from the Gulf Coast to Midwest refineries, and are also used to transport some refined products. Barge usage increased dramatically with the rapid development of the Bakken Shale (North Dakota and Montana) in the early 2010s, as crude oil could be shipped south by river to refineries in the lower Mississippi and Gulf Coast. In 2010, barges delivered 46 million barrels of crude oil

Oil barge at the Cape Cod Canal, 2011. Image credit: Pvalerio, Wikimedia Commons[11]

Petroleum and the Environment
Part 15: Transportation of Oil, Gas, and Refined Products

to U.S. refineries. This rose to 244 million barrels in 2014, before decreasing with the downturn in the industry, with 165 million barrels delivered in 2016.[12] Inland water transportation uses roughly 75% less energy than trucks and 25% less energy than rail, but is only viable where navigable rivers are close to both the source of oil and its destination.[10]

International Transportation

Crude oil, refined products, and natural gas move largely by pipeline between the U.S., Canada, and Mexico. In 2017, Mexico and Canada accounted for 44% of all U.S. petroleum imports[13] and 30% of all U.S. petroleum exports.[14]

Canada – the U.S. largely imports Canadian crude oil and exports refined products. Pipelines crossing the U.S.-Canada border include 31 oil and 39 natural gas pipelines, plus 16 pipelines moving other commodities such as carbon dioxide and industrial chemicals.[15] In 2016, 91% of all oil imported into the U.S. from Canada was transported by pipeline, 3% by rail, and 5% by marine vessels.[16]

Mexico – the U.S. largely imports Mexican crude oil and exports refined products, both by ship and by rail. The U.S. exports natural gas to Mexico via pipeline. U.S. natural gas exports to Mexico have grown from 333 billion cubic feet in 2010 to 1.37 trillion in 2016, and are projected to increase to 2.17 trillion cubic feet in 2020 as planned pipelines come online.[17]

Other countries – the U.S. exports and imports crude oil and refined products to and from other countries by sea. Total imports from countries other than Canada and Mexico were over 5 million barrels per day in 2017.[18] Historically, the U.S. has exported very little crude oil and a larger amount of refined products. However, exports of both have risen rapidly in recent years, and in 2017 total exports of crude oil and refined products to countries other than Canada and Mexico exceeded 4.4 million barrels per day.[19] Globally, marine vessels move huge volumes of crude oil, refined products, natural gas liquids, and natural gas around the globe. In 2016, oil, refined products, and natural gas represented roughly 30% of all international seaborne trade.[20]

In 2017, the U.S. exported over 650 billion cubic feet of natural gas as liquefied natural gas (LNG).[21] This represents a huge increase from about 50 billion cubic feet exported annually from 1985 through 2011, and is almost four times the amount of LNG exported in 2016. This increase was driven by increased domestic gas production, increased liquefaction capacity, and low-cost export facilities converted from import terminals that had been constructed to import LNG before the domestic gas boom began in 2006. Almost all the LNG is exported from the Gulf Coast, but as of early 2018, export terminals are being built at existing import terminals in Georgia and Maryland.[22] As of early 2018, almost half of all U.S. LNG exports were delivered to Mexico, South Korea, or China.[23] Natural gas produced in Cook Inlet, Alaska, was exported to Japan from Alaska's Kenai terminal in 2015 but not in 2016 or 2017,[24] pending decisions over the future of the terminal.

Storage

Storage facilities are used to ensure a steady supply of important fuels, reducing the risks of supply disruption due to natural disasters, conflicts, and other major events.

Natural gas and crude oil. The natural gas industry uses nearly 400 underground storage facilities to ensure that supply remains steady during periods of high demand.[25] The federal government's Strategic Petroleum Reserve can store over 700 million barrels of crude oil in underground salt caverns near refineries along the Gulf Coast (enough oil to supply the entire United States for roughly five weeks).[26] This reserve is part of a global storage system to respond to major supply disruptions, and was recently tapped in late August 2017 in response to supply disruptions from Hurricane Harvey.

Refined petroleum products. The northeast U.S. is home to two federal oil reserves for use in the event of a brief supply disruption. One holds a million barrels of home heating oil (created in 2000 and first used to reduce supply disruptions from Hurricane Sandy);[27] the other holds a million barrels of gasoline (created in 2014 in response to Hurricane Sandy).[28] Each reserve is spread over three locations in the Northeast. In total, these reserves contain roughly half a day's supply of gasoline and a day's supply of heating oil[29] (based on winter usage) for the entire northeast. However, supply disruptions rarely affect the entire northeast U.S., so shipments can be made from these reserves to targeted areas for a longer period of time.

Petroleum and the Environment
Part 15: Transportation of Oil, Gas, and Refined Products

Spills

A tiny proportion of all the oil transported in the U.S. is leaked or spilled. Most of the transportation system is safe and effective. However, because the quantities of liquids being transported are so enormous, even a tiny portion of this can be large enough to have significant environmental impacts in the area of the spill.

2017 spill volumes by mode of transportation:[30,31]
- **Rail** – 56,000 gallons of crude oil and refined products from seven reported incidents
- **Highway** – 180,000 gallons of crude oil and refined products from approximately 100 reported incidents
- **Pipelines** – 2.58 million gallons of crude oil and refined products from 328 reported incidents.

For each method of transport, spill volumes represent less than 0.001% of the total amount transported. The larger total amounts spilled from pipelines are mostly because almost all of the oil and refined products in the U.S. are transported by pipeline over long distances. Per barrel-mile, pipelines are less likely to spill than trucks (they don't get into crashes), but when they do spill, the spills can be larger and more difficult to clean up, especially because most pipelines are buried underground and carry large volumes of oil and refined products.

Although oil and refined product pipeline operators employ leak detection systems, such as pressure or flow monitoring and/or internal surveys of pipe integrity, roughly half of all leaks are found by the public or local operating personnel.[38]

The damaged portion of Enbridge Line 6B is prepared for removal, having spilled hundreds of thousands of gallons of heavy crude oil near Marshall, Michigan, in July, 2010. Image credit: U.S. Environmental Protection Agency.[32]

Transporting Water

Water used for drilling, oil and gas production, and hydraulic fracturing (not fully quantified,[33] but probably in the range of a few billion gallons per day[34]), and for refining oil (1-2 billion gallons per day)[35] must also be transported from its source to where it is needed. In addition, "produced water" extracted along with oil or gas (roughly 2.5 billion gallons per day)[36] must be transported to a disposal or reuse site. In established oil-producing areas or at permanent facilities, water may be moved by pipelines, but it is otherwise moved largely by truck. Oilfield waters are generally not federally regulated as hazardous wastes and so are not regularly measured and tracked. However, state or federal regulations may require certain management or handling practices for oilfield waters.[37]

For more information on natural gas leaks, see "Methane Emissions in the Oil and Gas Industry" and "Mitigating and Regulating Methane Emissions" in this series; for more information on spills, see "Spills in Oil and Natural Gas Fields".

Regulation of Transportation, Storage, Refining, and Marketing

The Pipeline and Hazardous Materials Safety Administration (PHMSA) in the U.S. Department of Transportation regulates the pipeline transport system and shipments of hazardous materials by land, sea, and air.[39] The Federal Energy Regulatory Commission (FERC) regulates the practices and rates of interstate oil pipelines,[40] while PHMSA regulates their operation. FERC also reviews applications for construction and operation of natural gas pipelines and LNG export and import terminals to certify their compliance with safety and environmental laws.

References & More Resources

For a complete listing of references, see the "References" section of the full publication, *Petroleum and the Environment*, or visit the online version at: www.americangeosciences.org/critical-issues/petroleum-environment

Oil Refining and Gas Processing
Turning complex mixtures into usable products

Introduction
Crude oil and natural gas are complex chemical mixtures that are generally unsuitable for direct use. Oil refining and gas processing turn these mixtures into a wide range of fuels and other products while removing low-value and polluting components.

Refining and processing have both positive and negative environmental impacts: although they remove harmful pollutants and produce cleaner-burning fuels, the operations at refineries and processing plants may release harmful pollutants into the environment, affecting local air and water quality.

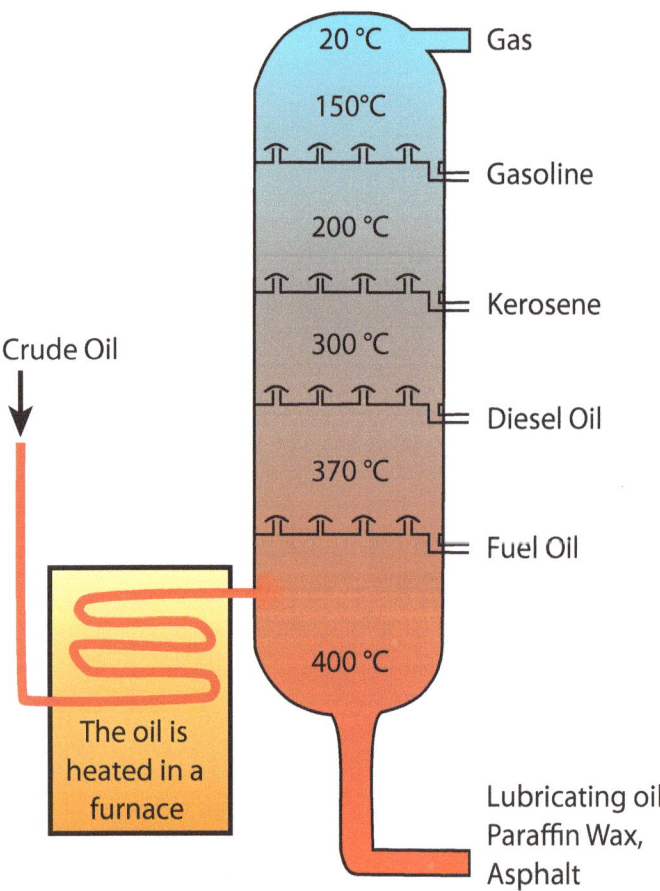

During crude oil distillation, different fuel types condense and are extracted at different temperatures. Image credit: Wikimedia Commons users Psarianos and Theresa Knott.[1]

Oil Refining
Crude oil is a mixture of many different hydrocarbon molecules of a range of sizes. Smaller molecules vaporize at lower temperatures, so crude oil can be **distilled** to separate out the different hydrocarbons. In the distillation process, crude oil is vaporized and the hot vapor rises up a column, cooling as it rises. Different hydrocarbons vaporize at different temperatures, so they condense into liquid form at different points in the column, separating the crude oil into different components that can then be further processed to optimize them for their final use.

Gasoline and diesel are the most lucrative products extracted from crude oil, so refineries use a range of techniques to maximize the production of these fuels. This may include **cracking** (breaking larger molecules down into smaller molecules[2]), **hydrotreating** (replacing impurities such as sulfur with hydrogen to improve fuel quality[3]), **reforming** (turning smaller molecules into gasoline[2]), **alkylation** (using an acid to produce high-octane gasoline from smaller molecules[4]), and **blending** (mixing different liquids together to produce uniform products that meet regulatory standards[5]). During the blending stage, ethanol from industrial ethanol plants is also blended into gasoline to increase its octane content, reduce carbon monoxide emissions, and meet the requirements of the Renewable Fuel Standard.[6]

Products of Oil Refining
Different crude oils have different compositions, containing different mixtures of hydrocarbons and variable amounts of sulfur and other impurities. The proportions of different refined products will vary with changes in the types of oil being refined, demand for different products, and regulations that influence this demand. Roughly 80-85% of all crude oil ends up as gasoline, diesel, or jet fuel. The rest is used to produce liquefied petroleum gases, petrochemical feedstocks, and a variety of other products.[7] In 2016, 141 U.S. refineries produced a daily average of 9.3 million barrels of gasoline, 3.7 million barrels of low-sulfur diesel, and 1.6 million barrels of jet fuel.[8]

Petroleum and the Environment
Part 16: Oil Refining and Gas Processing

Natural Gas Processing

In 2017, the United States produced 33 trillion cubic feet of natural gas.[9] A small fraction of this was used in field operations, re-injected into underground reservoirs, vented, or flared; the rest was processed by 550 gas processing plants to produce 27 trillion cubic feet of pipeline-quality natural gas.[10,11] Pipeline-quality gas must meet rigid standards for energy content and purity[12] for residential, commercial, and industrial use, including natural gas power plants.

Before processing, natural gas consists mostly of methane, with varying proportions of other hydrocarbons, carbon dioxide (CO_2), sulfur dioxide, nitrogen, water vapor, and helium.[13] Gas processing removes some of the non-methane components of natural gas in order to:
- Improve combustion and reduce corrosion by removing water
- Prevent the formation of damaging acids by removing harmful or corrosive gases – especially sulfur and CO_2 – that might otherwise react with small amounts of water to form acids
- Standardize the energy content of the gas to ensure uniform combustion in furnaces and other equipment, notably by removing non-combustible gases such as CO_2 and nitrogen
- Extract valuable minor gases for other uses (e.g., other hydrocarbons and helium)

Non-methane hydrocarbons extracted during gas processing are collectively called "natural gas liquids" (NGLs) because they form liquids more easily than methane at high pressure or low temperature. Of the NGLs, the most common are ethane, propane, and butane. Ethane and propane are further processed in large quantities to make feedstocks for plastics (see "Non-Fuel Products of Oil and Gas" in this series), while propane and butane are compressed into liquids to provide an energy-dense source of gas fuel for off-grid uses.

The main methods used to remove non-methane components from natural gas are absorbents and cooling. A variety of absorbents may be used, including special oils (for NGLs), glycol (for water), amines (for sulfur and CO_2,[14]), and zeolite or oil absorption (for nitrogen[15]). Chilling natural gas down to different temperatures allows different components to be removed as they condense into liquids. This is the most common method for nitrogen removal: the natural gas is chilled until the methane liquefies, allowing the nitrogen gas to be vented off.[16] NGLs may be removed in a single mixture that is then heated to different temperatures to isolate each NGL in turn.[18] After processing, the gas is deemed "dry" and ready to be shipped via pipelines to end users.

Oil refineries (open squares) and gas processing plants (blue) in the United States as of February 2018. Not shown: two refineries in Hawaii and five in Alaska. Image credit: U.S. Energy Information Administration.[17]

Refining, Processing, and the Environment

Refining and processing reduce the environmental impact of oil- and gas-derived fuels by removing harmful pollutants and improving their reliability during combustion. However, refineries and processing plants have their own environmental impacts, with corresponding procedures for minimizing those impacts. More information on these can be found in other parts of this series: "Mitigating and Regulating Methane Emissions" and "Air Quality Impacts of Oil and Gas."

Carbon dioxide (CO_2) occurs in varying proportions in natural gas and is removed at processing plants to improve the quality of the gas. Most of this CO_2 is vented to the atmosphere, accounting for roughly 0.4% of total U.S. greenhouse gas emissions (for comparison, methane leaks from the natural gas production and distribution chain are estimated to account for roughly 3% of U.S. emissions).[19] A small number of gas processing plants capture the CO_2 removed from natural gas during processing; this captured CO_2 is injected into oil fields to enhance oil recovery.[20]

References & More Resources
For a complete listing of references, see the "References" section of the full publication, *Petroleum and the Environment*, or visit the online version at: www.americangeosciences.org/critical-issues/petroleum-environment

Petroleum and the Environment
Part 17

Non-Fuel Products of Oil and Gas
Plastics, fertilizers, synthetic fibers, pharmaceuticals, detergents, and more

Introduction

Oil and natural gas are complex mixtures of chemicals. Oil refineries and gas processing plants extract the organic compounds that make the best fuels for transportation, heating, and electricity generation: gasoline, jet fuel, diesel fuel, heating oil, and methane. Other chemicals derived from processing oil and natural gas are called **petrochemicals** and are used to make thousands of non-fuel products.

Getting from Oil and Gas to Petrochemicals

Petrochemicals can be produced by refining oil or processing natural gas; petrochemical plants are typically built to use either oil- or gas-derived materials (or both), depending on the availability and price of each. Globally, most petrochemicals are derived from oil, but in the U.S., most petrochemicals are produced from natural gas, due largely to high domestic natural gas production.[1]

The most important raw materials for petrochemicals produced from natural gas are **ethane** and **propane**. After methane (which is mostly used for fuel), ethane and propane are the most common organic compounds in natural gas. They are removed during natural gas processing, so the more gas produced, the more ethane and propane are available to make petrochemicals.[2] Ethane and propane molecules are larger and heavier than methane, so they require less pressurization or cooling to turn them into liquids, earning them the name "natural gas liquids." During gas processing, other natural gas liquids are also derived in smaller amounts and are processed and used for many purposes, while methane itself is used as a source of hydrogen to make fertilizer and other products (see below). Naphtha, a liquid mixture distilled from crude oil at refineries, is also used as feedstock for various petrochemicals.[3]

Ethane, propane, or naphtha can be heated to very high temperatures (up to around 850 °C, or 1562 °F) to break apart molecules ("cracking")[4] or selectively pluck hydrogen atoms off the molecules ("dehydrogenation") to form new chemicals such as **ethylene** and **propylene**.

Ethylene and propylene are the two dominant petrochemicals: in 2016, the U.S. produced over 26 million tons of ethylene and over 14 million tons of propylene.[5] Ethylene is primarily converted into **polyethylene** (the most common plastic, used in thousands of applications), but is also used to make other plastics such as polyvinylchloride (PVC, for pipes and home siding) and polystyrene (used as a general plastic and as Styrofoam for insulation and packaging). Propylene is mostly converted into **polypropylene** for fibers, carpets, and hard plastic; some propylene produced during oil refining is used to make compounds that are added to gasoline to improve performance.[6] Both ethylene and propylene are used to make many other chemicals and materials with many uses, including specialty plastics, detergents, solvents, lubricants, pharmaceuticals, synthetic rubbers, and more.

From Petrochemicals to Consumer Products

Most **plastics**, **synthetic fibers** (such as polyester and nylon), and **resins** (such as epoxy) are produced from petrochemicals. These products are major components of vehicles, home and office buildings, electronics, clothing, packaging, and more.

Global Production of Major Petrochemicals (2016)

Ethylene (146 million metric tons):[5]
- Middle East: 19%
- United States: 18%
- China: 15%

Propylene (99 million metric tons):[5]
- China: 26%
- United States: 14%
- Western Europe: 13%

Ammonia (144 million metric tons of nitrogen):[7]
- China: 32%
- Russia: 9%
- India: 8%
- United States: 7%

Petroleum and the Environment
Part 17: Non-Fuel Products of Oil and Gas

While these are some of the best-known uses of petrochemicals, other major uses of petrochemicals and other non-fuel products of oil and gas include:

- **Fertilizers** – hydrogen derived from methane (the main ingredient in natural gas) is combined at high temperatures with nitrogen extracted from air to make almost all of the ammonia in the world (a small amount of ammonia is produced using other sources of hydrogen such as propane, naphtha, or gasified coal). About 88% of U.S. ammonia consumption is used as the nitrogen source for fertilizer. Other important uses of ammonia include household and industrial cleaning products, refrigerants, and in the manufacturing of plastics, dyes and explosives.[7]
- **Pharmaceuticals** – almost all pharmaceuticals are made from chemical feedstocks manufactured from petrochemicals and their derivatives.[8]
- Many **detergents** and other cleaning products are made from petrochemicals.[9] Similar cleaning products made from plant oils are now widely available, although these products are often also produced using substances made from petrochemicals.
- **Road asphalt** consists of roughly 95% crushed stone, sand, and gravel; the remaining 5% is a thick, dark oil known as asphalt or bitumen, which occurs naturally in some rocks but is also produced by oil refining.

Fertilizer: Ammonia from Natural Gas

Ammonia, a compound of hydrogen and nitrogen, is industrially produced on a vast scale. The U.S. produced over 10 million tons of ammonia in 2017,[7] almost entirely using hydrogen derived from natural gas. Roughly 88% of this ammonia is used to provide nitrogen for fertilizer, making up almost 60% of all fertilizer used in the United States.[10] With growing domestic natural gas production, low natural gas prices, and new fertilizer plants coming online,[11] U.S. ammonia production capacity is expected to grow by 25% from 2018 to 2022. Over this same period, global ammonia production (150 million tons in 2017) is expected to grow by 8%.[7]

Medicine: Plastics and Pharmaceuticals

Plastics are so widely used that it is easy to forget how varied they are – the most common plastic, polyethylene, comes in 10,000 different types for different uses.[12] In medicine, plastics serve a wide variety of purposes: keeping medical equipment

Appalachia's New Role in U.S. Manufacturing

Since 2010, the Appalachian region has seen enormous growth in natural gas production from the Marcellus and Utica shales: by 2017, they provided almost 25% of all U.S. gas production.[13] This gas is especially rich in ethane, propane, and other natural gas liquids,[14] providing vast resources for petrochemical production. By mid-2017, almost 23 billion cubic feet of natural gas was processed in Appalachia every day, providing 600,000 barrels per day of ethane and other natural gas liquids.[14] These natural gas liquids are mostly transported by pipelines to cracking plants on the Gulf Coast or exported to Canada.[15] However, the growth in natural gas liquid production, especially since 2013, has prompted plans to build new ethane crackers in Appalachia.[12] The first of these is likely to be a Shell project in Beaver County, Pennsylvania, which is due to begin operations in the early 2020s.[16] Local production of plastics and other petrochemicals is expected to support the local manufacturing industry.

Ammonium nitrate fertilizer being applied to winter wheat. Almost all ammonia used for fertilizer is derived from natural gas. Image credit: Michael Trolove, Wikimedia Commons.[17]

sterile; providing inexpensive disposable syringes, tubing, and single-use supplies to reduce the risk of infection; and forming implants and artificial joints, as well as many advanced materials, including natural-synthetic hybrids that can be used inside the body with lower risk of rejection.

Petroleum and the Environment
Part 17: Non-Fuel Products of Oil and Gas

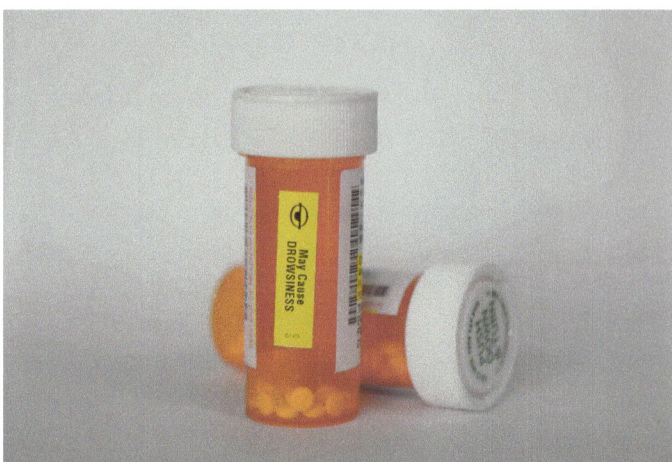

Medications and the bottles that house them are almost all made using petrochemical building blocks. Image credit: Airman Valerie Monroy, U.S. Air Force.[18]

One of the oldest uses of petrochemicals is petroleum jelly, a soft solid that often naturally separates from crude oil. Petroleum jelly was first marketed over 150 years ago and is still widely used for skin care and cosmetics.[19]

Petrochemicals provide the chemical building blocks for most medicinal drugs: nearly 99% of pharmaceutical feedstocks and reagents are derived in some way from petrochemicals.[8] For example, aspirin has been manufactured from benzene, produced in petroleum refining, since the late 19th century.[20]

Sulfur and Helium

Sulfur is common in both oil and natural gas. If it is not removed, it can corrode steel transportation equipment and pipelines, and produce acid rain when released as sulfur dioxide during combustion. Removing sulfur from these fuels reduces economic and environmental damage while also producing a valuable industrial material: in 2017, the United States produced over 9 million tons of sulfur valued at $585 million; the vast majority of this came from oil refining and gas processing.[21] Sulfur is mostly used to make sulfuric acid for a wide range of industrial processes, notably in the production of fertilizer, which accounts for half of global sulfur consumption.[22]

Helium is an important industrial gas with a wide variety of applications in aircraft, aerospace, electronics, and advanced metalworking. Liquid helium is the coldest cryogenic liquid available: aside from advanced research, a major application of liquid helium is to cool magnetic resonance imaging (MRI) scanners. Helium is sourced entirely from natural gas at gas processing plants. The U.S. is the largest helium producer in the world: in 2017, the U.S. accounted for 57% of global helium production. Qatar produces most of the rest (28% of global production), with a handful of other countries producing small quantities.[23]

Petrochemicals and the Environment

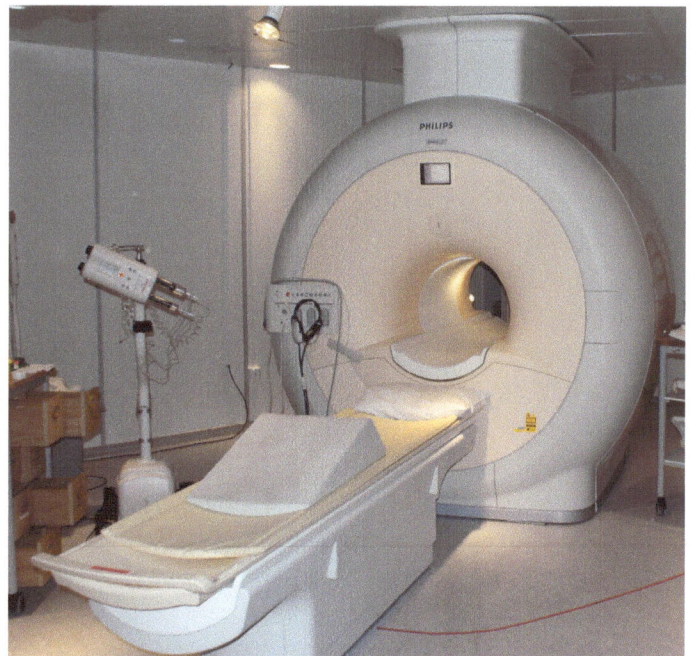

Magnetic resonance imaging (MRI) scanners use specialized magnets that require extremely low temperatures to operate. These low temperatures are achieved using liquid helium, which is the only currently available source of ultra-low cryogenic temperatures. Helium is sourced entirely from natural gas. Image credit: Jan Ainali, Wikimedia Commons.[24]

When oil, gas, and the environment are discussed together, the focus is often on the ways in which oil and gas can damage environmental or human health, and steps that can be taken to prevent or reduce these impacts. Petrochemicals are an interesting case because they have both positive and negative impacts on the environment. The negative impacts are significant and should not be understated, but they are also generally well-known: the accumulation of plastic waste,[25] the harm that plastics and their breakdown products can cause when ingested by animals,[26] and the damage to aquatic ecosystems caused by fertilizer runoff (caused by both natural and petrochemical

Petroleum and the Environment
Part 17: Non-Fuel Products of Oil and Gas

fertilizers).[27] The environmentally beneficial uses of petrochemicals are less commonly discussed but are an important part of any discussion of overall environmental impacts. Some of these environmentally beneficial uses include:

- Oil absorbents used to clean up oil spills.[28] Natural biological materials (e.g., feathers or straw) can absorb 3-15 times their weight in oil, while minerals (e.g., sand and vermiculite clay) can absorb 4-20 times their weight in oil. These natural materials are inexpensive but they can often sink in water, making them less effective for oil spills at sea. Synthetic oil absorbents made from plastic or nylon can absorb up to 70 times their weight in oil while staying afloat, making them effective for oil spill cleanup in water.
- Federal regulations require construction or demolition sites to have barriers that prevent stormwater runoff from carrying sediment and waste into sewers, rivers, or streams.[29] These barriers are often made of plastics or synthetic fibers, which are weather-resistant and impermeable.
- Some petrochemical-derived products can be used to remove carbon dioxide from the combustion gases of power plants, thus reducing the greenhouse gas emissions of energy production. In 2017, the Petra Nova project near Houston, Texas, began operations as the largest such project in the world. Attached to an existing coal-fired power plant, the system removes carbon dioxide from the combustion gases by reacting it with chemicals called amines (derived from ammonia, which is produced from natural gas).[31] The Petra Nova project is designed to capture 1.6 million tons of carbon dioxide per year.[32]

Plastic fencing to control soil erosion and stormwater runoff at a construction site. Image Credit: U.S. Environmental Protection Agency.[30]

Plastic Disposal and Recycling

Most plastic food and beverage containers, plastic bags, and other packaging can be recycled.[33,34] Some plastics can be reprocessed to make the same product they were originally used for (e.g., plastic bottles). Plastic bottles can also be turned into polyester fiber used for fleece jackets, insulation, and carpeting. Plastic bags and film can be recycled into plastic lumber, used to make outdoor furniture, decks, and fencing.[35]

Despite the many potential reuse options for plastics, only about 9.5% of plastic material generated in the U.S. was recycled in 2014. Over 75% went into landfills, while 15% went into trash-to-energy plants that burn waste to generate electricity.[36] A small but significant proportion of used plastic is not properly disposed of and ends up in the surface or marine environment, where it takes hundreds of years to decompose and can harm wildlife.

References & More Resources

For a complete listing of references, see the "References" section of the full publication, *Petroleum and the Environment*, or visit the online version at: www.americangeosciences.org/critical-issues/petroleum-environment

Penn State Extension – Gas, Cracker, Polymer, Pellets – Ethane's Journey to Plastics. Shale Gas Webinar, September 21, 2017. https://extension.psu.edu/gas-cracker-polymer-pellets-ethane-s-journey-to-plastics

U.S. Geological Survey – Mineral Commodity Summaries. (Information on commodities derived from oil and gas and related operations, including sulfur, helium, ammonia, and iodine). https://minerals.usgs.gov/minerals/pubs/mcs/

Hess, J. et al. (2011). Petroleum and Health Care: Evaluating and Managing Health Care's Vulnerability to Petroleum Supply Shifts. Am. J. Pub. Health, 101(9), 1568-1579. https://ajph.aphapublications.org/doi/10.2105/AJPH.2011.300233

Petroleum and the Environment
Part 18

Air Quality Impacts of Oil and Gas
Emissions from production, processing, refining, and use

Introduction

All widely used combustible fuels emit harmful (toxic or ozone-forming) gases and particles when burned to provide energy. These air pollutants can have a wide array of public health impacts, such as increasing the rate of certain cardiovascular (heart) and pulmonary (lung) diseases, cancers, and strokes.[1] Mobile emission sources, including cars, trucks, trains, and airplanes using gasoline, diesel, and aviation fuel account for more than half of all air pollutant emissions in the United States.[2] Stationary sources such as oil- and gas-fields, oil refineries, and gas processing plants are small in comparison to those from burning fuels but may have significant local impacts.

Oil and gas production, processing, and use also release large quantities of greenhouse gases, especially methane and carbon dioxide. Although these gases have climatic impacts and are regulated by the U.S. Environmental Protection Agency (EPA), they do not significantly impact local air quality except at high concentrations in confined spaces, so they are not discussed in this sheet. For more information on methane and carbon dioxide emissions, see other sections in this series: "Petroleum and the Environment: An Introduction", "Methane Emissions in the Oil and Gas Industry", "Mitigating and Regulating Methane Emissions", and "Oil Refining and Gas Processing".

Major Air Pollutants and the Regulatory Framework

Six major pollutants produced directly or indirectly by burning fossil fuels (in addition to other human activities) are required by law to be regulated by the EPA under the Clean Air Act: **carbon monoxide** (CO), **nitrogen oxides** (NO_x), **sulfur oxides** (SO_x) **ozone** (O_3), **particulate matter** (PM), and **lead** (Pb).[4]

The EPA and states also set standards for 187 other "**air toxics**" with serious known or suspected health effects.[5] These include some oil and gas components, such as benzene, which is found in gasoline. Oil- and gas-related air toxics may be emitted during petroleum exploration, production, refining and processing, and combustion. The wide range of potential emission sources requires a variety of regulations to effectively manage different emissions at different points in the oil and gas supply chain.

The Environmental Protection Agency (EPA) takes two general approaches to regulating air pollution related to petroleum exploration and production activities:

- The EPA sets ambient air quality standards for the six major air pollutants mentioned above, and periodically revises these standards based on current scientific information about health impacts and the availability of pollution-reducing technology.[6] States develop and enforce implementation plans to control air pollutants, and tailor these plans to their specific sources to comply with EPA air quality standards. Areas within a state that fail to meet air quality standards ("non-attainment areas") face additional restrictions.[7] State rules may regulate vehicle emissions and the construction and operation of industrial plants including petroleum refining, storage, transportation, and marketing facilities.

Flaring of unprocessed natural gas, as seen here on a well site in North Dakota (flare on right of picture), is one of several oilfield processes that can negatively affect air quality, as well as increasing carbon dioxide emissions. Flaring in North Dakota was particularly common in the early 2010s, as oil production was booming but pipelines were not in place to collect and transport the natural gas produced from oil wells. Image credit: Tim Evanson, Flickr.[3]

Petroleum and the Environment
Part 18: Air Quality Impacts of Oil and Gas

- The EPA also sets maximum legal pollutant emissions and provides guidelines on the appropriate type of pollution control for the six major pollutants and specific air toxics at certain stationary sources across industries.[8] Those related to oil and gas include gas turbines, boilers, gasoline terminals, and petroleum refineries.

New emission regulations are often contentious. New regulations may require industries to install new equipment or adopt new procedures to ensure compliance, sometimes at significant cost; the health or safety benefits of a new regulation may be debated; and some may argue that a regulation does not do enough to meet environmental or public health needs. Regulators receive public input from a wide variety of stakeholders while developing new regulations. However, because stakeholders often have many competing priorities, the final regulations are often challenged in court as being either too mild or too onerous.

Ozone and Oil & Gas Production

Oil- and gas-producing areas may have high levels of volatile organic compounds (VOC) that contribute to harmful ozone formation in the lower atmosphere. VOCs are emitted by vehicles and equipment used in oil operations as well as in surrounding roads and communities. VOCs also evaporate directly from the oil and gas being extracted, stored, and transported around the oilfield.[9] Storage tanks, certain types of pumps and compressors, and leaky valves and fittings may allow VOCs to escape into the atmosphere. High levels of VOCs have been measured in some oil- and gas-producing areas, such as the Eagle Ford in South Texas.[10,11] However, quantifying these oilfield emissions and their effects on air quality is often complicated by variable and/or unknown background ozone levels; determining the contribution of oilfield emissions to ambient VOC levels is an active area of current research.

Air Pollution from Oil Refineries and Gas Processing Plants

Oil refineries convert crude oil into a vast array of products, requiring a range of operations that emit a variety of pollutants, both through normal operations and in accidents. The EPA's enforcement of the Clean Air Act at petroleum refineries has led to dozens of settlements in which companies have agreed to invest billions of dollars in control technologies and supplemental environmental projects.[12] Due largely to this enforcement, from 2000 to 2011, combined SO_x and NO_x emissions from refineries decreased by 75%.[13] Other regulations control VOC and benzene emissions from equipment at refineries and petroleum product distribution systems.[14]

Natural gas processing has fewer air pollution risks, but some plants that process natural gas containing hydrogen sulfide – a poisonous, corrosive, and flammable gas – are regulated as toxic release sites.[15] The range of refinery and processing plant operations and their related emissions are summarized in the EPA's Compilation of Air Pollutant Emission Factors.[16]

Refinery emissions can be reduced by many means, including improved furnace efficiency, leak detection and control, gas recycling, reduced flaring and venting, scrubbers, and catalytic reactors.[17,18,19,20] From 1990 to 2013, refinery emissions of air toxics decreased by 66% and VOC emissions decreased by 69%, despite a 14% increase in the amount of crude oil being processed.[21] For communities near refineries, such as Galena Park near Houston, Texas (see figure), sustained future decreases in emissions from refineries and fuel storage facilities are an important part of mitigating the local public health impacts of oil and gas activity.[22]

Refineries, fuel shipping terminals, and other industrial facilities along the Houston Ship Channel, 1999. The city of Galena Park is center-right. Roughly half of all U.S. refining capacity is located on the Texas and Louisiana coasts, leading to a higher risk of air pollution from refineries in this region of the Gulf Coast. Image credit: Army Corps of Engineers.[23]

Petroleum and the Environment
Part 18: Air Quality Impacts of Oil and Gas

Combustion and U.S. Air Quality Over Time

In the United States, attempts to decrease the emission of pollutants from fossil fuel combustion have been highly successful, especially since the Clean Air Act of 1970 (with revisions in 1977 and 1990).[4] From 2000-2016, average air concentrations decreased for ozone (by 17%), particulate matter (42%), lead (93%), sulfur dioxide (72%), nitrogen dioxide (47%), and carbon monoxide (61%).[24] Reflecting this improvement, the total number of days per year that 35 major metropolitan areas reported their air quality index as "unhealthy for sensitive populations" or worse decreased from 2,076 days in 2000 to 697 days in 2016, a reduction of 66%.[25]

These pollutants have been reduced in a number of different ways:

- **Carbon monoxide** and **NO_x** emissions have been reduced by fitting engine exhausts with catalytic converters (to convert carbon monoxide to carbon dioxide) and NOx traps (to remove NO_x).

- **Lead** was added to gasoline starting in the mid-1920s to improve engine operation, but it is a potent neurotoxin. Atmospheric lead levels in the U.S. began to decrease in the 1970s with EPA-mandated reductions in leaded gasoline and the introduction of catalytic converters to car exhausts, which require unleaded gasoline. Since 1996, lead has been completely banned from all gasoline for road vehicles in the United States,[26] although as of 2018 some small aircraft are still permitted to use aviation gasoline containing lead.[27] Since 1990, average atmospheric lead concentrations in the U.S. have decreased by 99%.[28]

- **Sulfur oxides** (SO_x, mostly sulfur dioxide, SO_2) are primarily produced by the burning of fossil fuels (especially coal) in power plants and other industrial facilities. Sulfur emission reductions started with ambient air quality standards set by the EPA in the 1970s,[29] and continued with the 1990s Acid Rain Program, which set procedures for coal-fired power plants to reduce SO_x and NO_x emissions that contribute

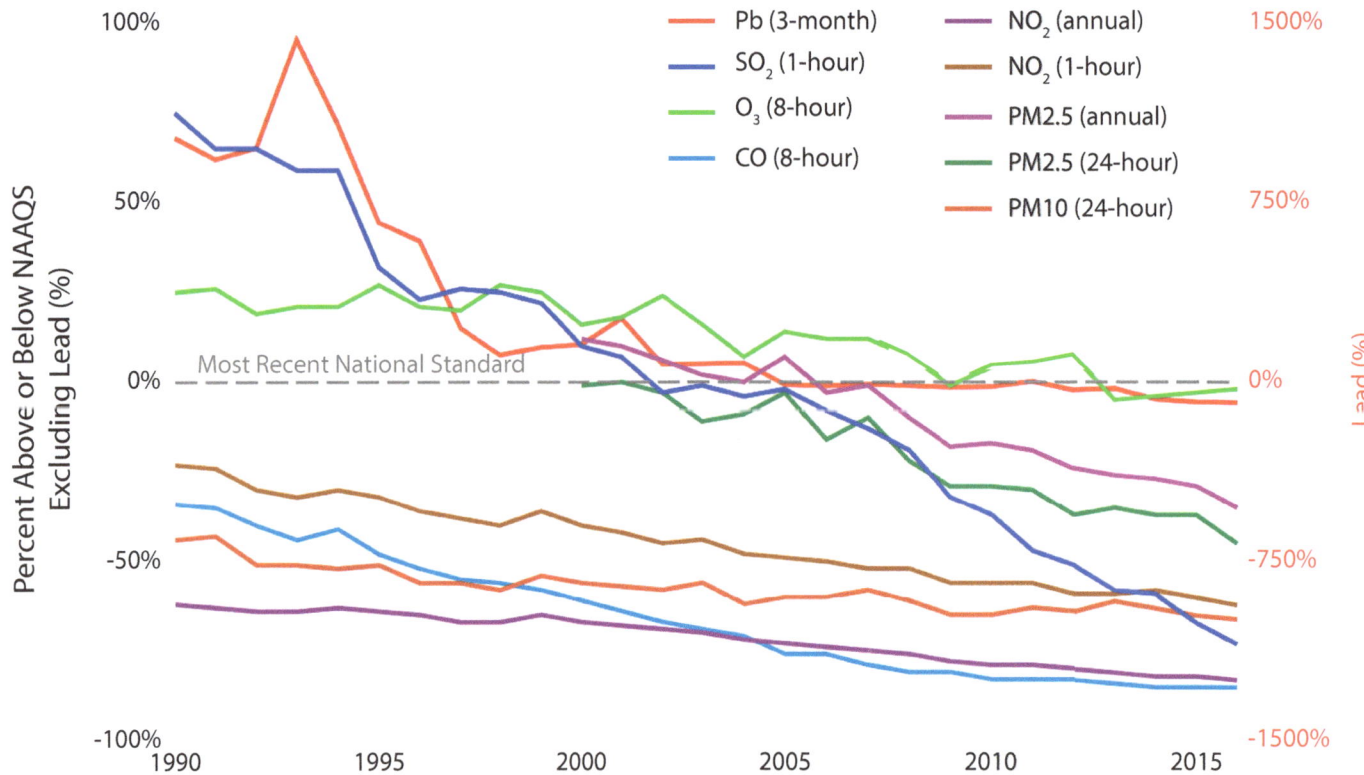

Changes in the average U.S. concentrations of the EPA's six target air pollutants from 1990-2016. Time periods indicate duration over which measurements are averaged. NAAQS = National Ambient Air Quality Standards. Image credit: U.S. Environmental Protection Agency.[31]

Petroleum and the Environment
Part 18: Air Quality Impacts of Oil and Gas

to acid rain.[30] The switch from coal to natural gas power generation since the 1980s has further cut SO_x emissions. Refineries and processing plants are also required to capture sulfur from oil and gas, reducing SO_x emissions upon combustion.

Particulate matter and **ozone** (major ingredients of smog) are both produced by the reaction of other pollutants in the atmosphere. In addition, particulate matter is emitted directly in gasoline- and diesel-vehicle exhaust. In the U.S., the EPA has set ambient air quality standards for particulate matter and ozone since 1970, reviewing and tightening these standards on several occasions.[32,33] Enforcement of EPA air quality standards decreased total emissions of particulate matter from vehicles by more than 50% from 2002 to 2016.[34] Standards issued in 2007 and 2014 will continue to lower emissions as the vehicle fleet turns over to newer vehicles that conform to these standards.[35] The story is similar for air toxics: from 1990 to 2014, average atmospheric concentrations of air toxics decreased by 68%,[24] and from 1990 to 2009, benzene levels decreased by 66%.[36]

Despite these improvements, air pollution still presents a serious health risk in urban areas. The EPA estimates that 14 million people in roughly 60 urban areas of the U.S. have more than a 1-in-10,000 lifetime risk of developing cancer caused by air pollution – ten times higher than the overall U.S. population.[38] Over time, the past success of regulations and technologies in improving U.S. air quality means that additional emissions reductions come at increasing cost relative to the size of the reduction.[39] Evaluating the economic, environmental, and health impacts, the difficulty of measuring and understanding impacts on large and diverse human populations, and the huge variety of different stationary and mobile emission sources combine to make air quality management a perennially controversial subject.

Smog visible as an orange haze over Los Angeles. Large urban areas in the United States still experience degraded air quality due in large part to the emission of smog-forming molecules and particulate matter from gasoline and diesel engines in vehicles. Image credit: Al Pavangkanan, Flickr.[37]

References & More Resources

For a complete listing of references, see the "References" section of the full publication, *Petroleum and the Environment*, or visit the online version at: www.americangeosciences.org/critical-issues/petroleum-environment

World Health Organization – Ambient (outdoor) air quality and health. http://www.who.int/mediacentre/factsheets/fs313/en/

U.S. Environmental Protection Agency – Air Emissions Sources. https://www.epa.gov/air-emissions-inventories/air-emissions-sources

National Institute of Environmental Health Sciences – Air Pollution. https://www.niehs.nih.gov/health/topics/agents/air-pollution/index.cfm

U.S. Environmental Protection Agency – Air Quality – National Summary. https://www.epa.gov/air-trends/air-quality-national-summary

U.S. Environmental Protection Agency – Benefits and Costs of the Clean Air Act 1990-2020, the Second Prospective Study. https://www.epa.gov/clean-air-act-overview/benefits-and-costs-clean-air-act-1990-2020-second-prospective-study

Petroleum and the Environment
Part 19

Methane Emissions in the Oil and Gas Industry
Quantifying emissions and distinguishing between different methane sources

Introduction
Methane is the main component of natural gas, a cheap, abundant, and versatile source of energy that produces less carbon dioxide than other fossil fuels when burned. However, methane itself is a more potent greenhouse gas than carbon dioxide. Methane leaks from wells, pipelines, or processing equipment can substantially increase the greenhouse gas emissions of the natural gas sector, while also wasting resources as methane escapes into the atmosphere.

Identifying Methane Sources
Methane may be produced in two ways. Thermogenic methane, the source of most natural gas reserves, is produced by the effects of heat and pressure on the deeply buried remains of marine microorganisms, and usually occurs with oil. Biogenic methane is produced by microbes in the stomachs of cows, sheep, goats, and other ruminant animals (known as enteric fermentation), and in manure, shallow coal and oil deposits, and wetlands. Identifying whether a methane source is thermogenic or biogenic is crucial for determining the methane emissions from oil and gas operations. This section of Petroleum and the Environment focuses on quantifying emissions of methane into the atmosphere; other parts of this series cover efforts to reduce methane emissions ("Mitigating and Regulating Methane Emissions"), and issues of methane in groundwater ("Groundwater Protection in Oil and Gas Production").

U.S. Methane Emissions
Determining the relative methane emissions from different sources is very difficult. The majority of methane emissions come from several vast industries that often operate right next to each other (agriculture, oil and gas, mining, and waste management). Leaks can be short-lived or prolonged, and emission rates from agriculture and landfills change over time. So although atmospheric methane levels can be measured very accurately, there is a great deal of uncertainty in the overall proportion of emissions coming

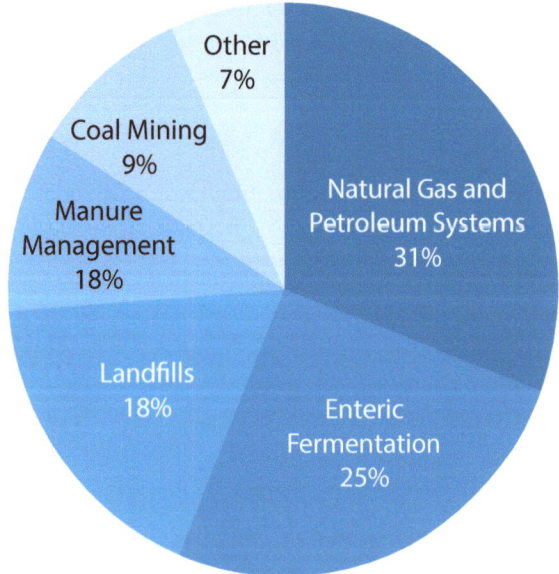

EPA estimates of U.S. methane emission sources in 2015. Image credit: American Geosciences Institute, modified from the U.S. Environmental Protection Agency.[1]

Methane Facts and Figures

- In 2015, methane made up about 10% of U.S. greenhouse gas emissions in terms of global warming potential; carbon dioxide (CO_2) made up 82%.[1]
- Natural gas provided 31.5% of U.S. electricity in 2017 – the largest single source of electricity in the country.[2]
- Natural gas power generation produces 50-60% less CO_2 than coal to produce the same amount of energy,[3] but methane leaks reduce this emissions-saving benefit.
- EPA estimates of methane emissions from natural gas systems decreased by 16% from 1990 to 2015. EPA-estimated methane emissions from crude oil and refined oil product systems decreased 28% from 1990 to 2015.[4] However, emissions estimates remain uncertain.
- In addition to livestock, manure, mining, and landfills, other major sources of global methane emissions also include wetlands and rice paddies.[5]

Petroleum and the Environment
Part 19: Methane Emissions in the Oil and Gas Industry

from different human activities. The national numbers in this sheet are best available estimates but may not be fully accurate.

Since the early 1990s, the U.S. Environmental Protection Agency (EPA) has annually released the U.S. Greenhouse Gas Inventory[4] as part of U.S. reporting to the United Nations in accordance with the Framework Convention on Climate Change.[6] The inventory is based on emissions reports from more than 8,000 industrial, manufacturing, and oil and gas facilities; power plants; and landfills.[7] These reports represent only about half of all U.S. greenhouse gas emissions, resulting in large uncertainties in emissions volumes.

Emissions from Oil and Natural Gas Systems

The oil and natural gas system is one of the most complex sources for emissions estimates because of the number of emission sources, their technical complexity, and the variability between different facilities.[8,9] Similar facilities may report different emissions,[8] and emission volumes may change over time as new leaks arise and are detected and repaired.[10]

Reflecting this complexity, the EPA estimate of the overall methane leak rate from the U.S. natural gas system has changed over time as new information has become available.[11] For example, between 2010 and 2011, the EPA's leak estimate for the year 2008 was updated from 96 to 212 million metric tons of carbon dioxide-equivalent; in 2013 this was then revised down to 163 million metric tons.[13] Estimates have not varied as widely from 2014 to 2017, but there remains considerable uncertainty in these figures.

Regional Emissions Studies

Detailed studies of major oil- and gas-producing areas can identify biogenic vs. thermogenic methane sources, monitor smaller sources not included in the EPA inventory, and identify particularly leaky equipment. Location-specific studies have been a major research focus in recent years.[14] For example:

- A study of seven oil- and gas-producing regions in the U.S. found higher methane emissions in mainly oil-producing areas than in mainly gas-producing areas. This in part reflects the fact that oil may contain some methane that can escape from oil storage tank vents and other openings.[15]

- In the Barnett shale area around Dallas and Fort Worth, Texas, 67% of methane emissions are from oil and gas sources.[16] Half of all oil and gas methane emissions in this area come from just 2% of production, processing, and transportation facilities, and 90% of emissions come from just 10% of facilities.[17] This suggests that most of the natural gas infrastructure is reliable, but a small number of "super-emitting" sites have major leaks. Super-emitting sites are expected to change over time as equipment accrues damage and is repaired or replaced. Detecting and reducing emissions therefore requires continuous monitoring.[10]

The extent of methane leaks from the natural gas system is one of the largest uncertainties regarding the environmental impact of the oil and gas industry. Working towards a comprehensive understanding of methane emissions is a major area of ongoing research, involving a combination of large-scale regional measurements and focused local studies from the ground, air, and space.

Improvements in remote sensing technologies are allowing increasingly high-precision measurements of regional methane emissions from plane-mounted sensors and even satellites. MethaneSAT (artist's impression pictured), a partnership led by the Environmental Defense Fund and launching in 2020 or 2021, will measure methane emissions from fifty major oil- and gas-producing regions around the world. Image credit: Environmental Defense Fund.[12]

References & More Resources

For a complete listing of references, see the "References" section of the full publication, *Petroleum and the Environment*, or visit the online version at: www.americangeosciences.org/critical-issues/petroleum-environment

Petroleum and the Environment
Part 20

Mitigating and Regulating Methane Emissions
Efforts to decrease emissions from the oil and gas industry

Introduction
Methane is the main component of almost all natural gas, and gas delivered to end-users is purified to 95-98% methane.[1] There are three main sources of methane emissions from the oil and gas industry:
- When a well is being drilled, cleaned out, or hydraulically fractured. As the fluids involved in these operations flow back up the well to the surface they bring methane with them, which can escape into the atmosphere.
- Through leaky equipment in wells, processing plants, and pipelines.

Uncaptured natural gas being flared off in the Bakken oil field, converting the methane to carbon dioxide. Image credit: Wikimedia Commons user Joshua Doubek.[2]

- When operators burn off ("flare") or vent small amounts of methane produced from oil wells. This commonly occurs when the gas cannot be sold due to low quality or lack of pipeline access. Flaring converts methane to carbon dioxide, but venting directly emits methane into the atmosphere.

There is a strong environmental incentive to reduce emissions of methane: it is a potent greenhouse gas that traps much more heat per molecule than carbon dioxide.[3] Methane in natural gas also coexists with a number of other organic compounds that contribute to the formation of ozone, which is harmful to plants and animals, including humans.[4] Some efforts to reduce these kinds of emissions are voluntary;[5] others are legally required by government regulations. Some regulations are enacted at the federal (national) level by the Environmental Protection Agency (EPA),[6] while others are enacted and enforced by individual states.[7] The EPA also supports voluntary programs to assist both the energy and agricultural industries in reducing their methane emissions.

Methane-related regulations include the use of "green" (reduced-emission) completions for gas wells, reduction of gas flaring, and leak monitoring and repair.

Green Completions of Oil and Gas Wells
Well "completion" involves all of the processes needed to get a well ready to produce oil and/or gas, including hydraulic fracturing. During these processes, mud and water from drilling and hydraulic fracturing (if used) flow back up the well to the surface along with some water contained in the oil- or gas-bearing rocks. This stream of fluids often brings oil and gas with it, and the gas has historically been allowed to escape into the atmosphere or flared off (see below). "Green" completions use specialized equipment to capture these gases and fluids.[8] Captured methane may then be used on-site or sold.

Since 2015, the EPA has required that green completions be used for all hydraulically fractured natural gas wells.[9] Prior to this, Wyoming,

Petroleum and the Environment
Part 20: Mitigating and Regulating Methane Emissions

Colorado, and some cities in Texas had each implemented their own regulations requiring green completions, and some operators had voluntarily used green completions in other areas.[9] As of 2018, the EPA regulations requiring green completions for oil wells are being considered for possible revisions.[10]

Reducing Natural Gas Flaring

In areas that produce mostly oil with small amounts of gas, operators often start moving the high-value oil to refineries before pipelines can be built to transport the less valuable natural gas. Instead, this gas is simply "flared," i.e., burned in an open flame. This controlled burning prevents the buildup of flammable methane and converts methane to carbon dioxide – a less potent greenhouse gas. However, flaring wastes usable energy, increases the carbon footprint of the industry, and decreases the royalties that landowners – including private citizens and the government – could earn from the sale of this methane.

Flaring is typically most common in new areas of high oil production, and decreases as pipelines are built to transport the gas. For example, flaring was widely used in the rapidly growing Bakken area of North Dakota in the early-mid 2010s. From early 2014 to early 2016, flaring in North Dakota fell from 36% to 10% of produced gas as production fell (due to lower oil prices) and a local network of pipelines was constructed to collect the gas from individual wells.[11]

Natural gas infrastructure includes a large number of valves and connectors, such as at this gas processing facility in Montgomery County, Texas. Components such as these may be sources of gas leaks. Image credit: Roy Luck, Flickr.[12]

The Bureau of Land Management (BLM) issued a rule in November 2016 that limited flaring on federal lands.[13] In late 2017, parts of this rule were temporarily suspended or delayed until January 2019, pending review of the rule by BLM.

Leak Detection and Mitigation

In 2016, the EPA issued emissions standards aimed at reducing methane emissions from new sources or facilities by detecting leaks and repairing or replacing leaking equipment at oil and natural gas wells, processing plants, pipelines, gas compressors, and valves and connectors in these systems.[14] As of 2018, the Administration, Congress, and the courts are involved in multiple ongoing actions to amend, advance, or delay the regulation.[10,15]

Some states, such as California[16] and Pennsylvania,[17] require operators to perform periodic leak detection surveys of oil and gas facilities, followed by mandatory repair or replacement if leaks are found.

Voluntary Emissions Reduction Programs

The EPA's voluntary Natural Gas STAR certification program, which includes over 150 partnerships with oil and gas production, transmission, and distribution companies, encourages and documents the use of emissions-reducing technologies.[18]

Because methane emissions are also a major issue in agriculture, EPA's AgSTAR program promotes the use of biogas recovery systems to capture methane emissions from livestock waste to be used as fuel.[19]

References & More Resources

For a complete listing of references, see the "References" section of the full publication, *Petroleum and the Environment*, or visit the online version at: www.americangeosciences.org/critical-issues/petroleum-environment

Bradbury, J. et al. (2013). Clearing the Air: Reducing Upstream Greenhouse Emissions from U.S. Natural Gas Systems. Working Paper. Washington, DC: World Resources Institute. http://www.wri.org/publication/clearing-air

U.S. Regulation of Oil and Gas Operations
Federal and state regulation of exploration, production, transportation, and more

Overview

Regulation of oil and gas operations has existed in various forms for over 100 years.[1] Regulation has several objectives: protecting the environment (including air and water quality), protecting cultural resources, protecting workers' and the public's health and safety, and reducing wasted resources.[2,3,4]

Federal, state, and local governments each regulate various aspects of oil and gas operations. Who regulates what depends on land ownership and whether federal regulations or state laws apply. In general, most drilling and production is regulated by the states. Federal regulations primarily safeguard water and air quality and worker safety, as well as exploration and production on Native American lands, federal lands, and the Outer Continental Shelf.

Regulations are implemented by the executive branches of local, state, and federal government based on the laws enacted by local, state, and federal legislators. Public input is a formal part of regulation development. The Clean Air Act (1963), the Clean Water Act (1972), and the Safe Drinking Water Act (1974), plus later revisions to these laws, form the basis of most federal regulation of the oil and gas industry. State roles in regulating oil and gas drilling and production were formalized by the Interstate Oil & Gas Compact Commission (IOGCC), which formed in 1935 to set standards for oil and gas drilling and develop production regulations that the states agreed to enact.[7]

State Regulation of Exploration and Production[8]

Exploration and production on state and private land are regulated by each of the 33 oil- and gas-producing states. States also regulate all oil and gas operations in state waters that extend from the coast to 3 to 9 nautical miles from the shoreline, depending on the state. Local zoning may control

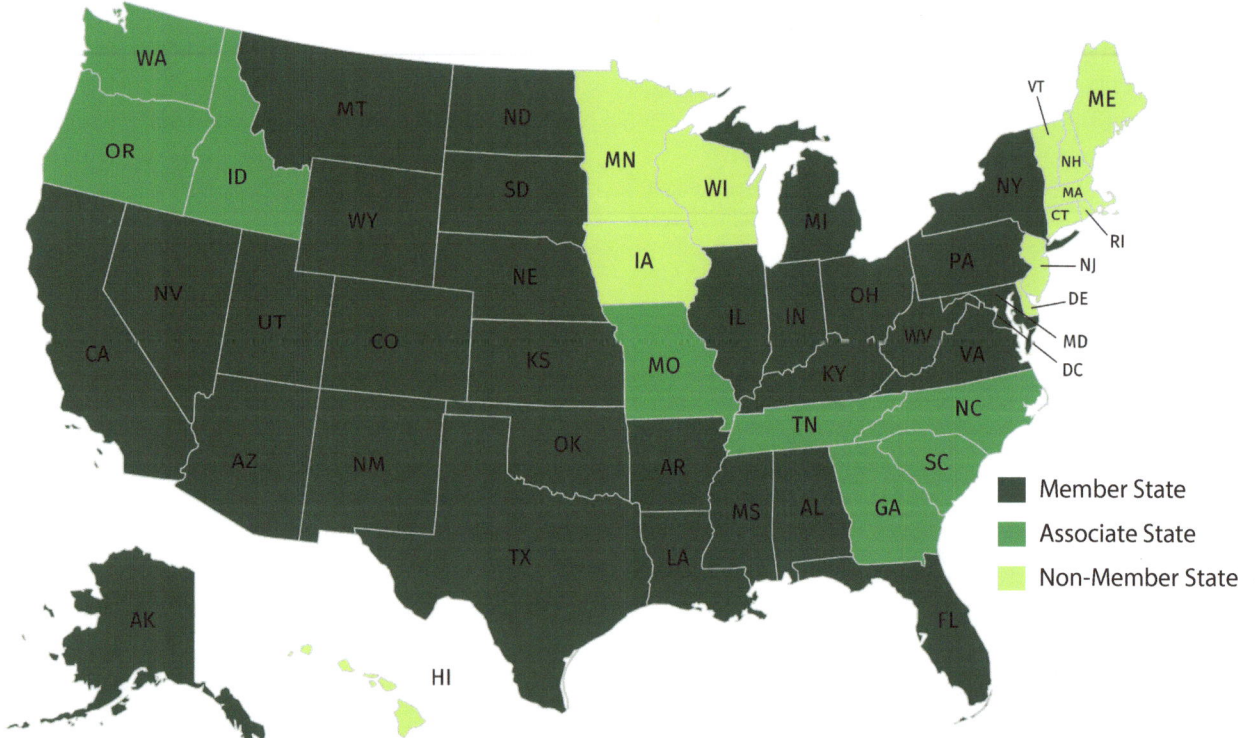

States belonging to the Interstate Oil & Gas Compact Commission.[5] Image credit: American Geosciences Institute, produced with mapchart.net.[6]

Petroleum and the Environment
Part 21: U.S. Regulation of Oil and Gas Operations

some activities such as the minimum distance wells and other facilities must be set back from homes and businesses.

State regulations vary from state to state and over time. Early state regulations were largely focused on preventing waste, ensuring the rights of mineral owners to develop their resources, and conserving resources to ensure the viability of future production. Environmentally focused regulations have become increasingly prominent over time, especially since the 1970s.[9] State-regulated activities include seismic and other geophysical surveys, leasing, drilling, hydraulic fracturing, oil and gas production, well closure, and site restoration. States enforce their regulations through permitting and regulatory inspections.

In fiscal year 2016, production from federal and Native American lands, onshore and in the Outer Continental Shelf, was 23% and 17% of total U.S. production for oil and gas, respectively.[10,11,12]

Federal Regulation of Exploration and Production

On Non-Federal Land

The federal role in regulating exploration and production primarily focuses on environmental protection. The Environmental Protection Agency (EPA) sets standards on drinking water and air quality under the authority of the Clean Air Act, the Clean Water Act, and the Safe Drinking Water Act.[14] In most cases, the EPA allows states to develop and implement the regulations necessary to meet federal standards. In a few areas, the EPA's regulatory role is more direct:

- The EPA requires the capture of all gases and fluids that come out of a well as it is being prepared for oil/gas production, including during hydraulic fracturing. This "green completion" rule, required for all natural gas wells since January 2015, requires equipment and procedures

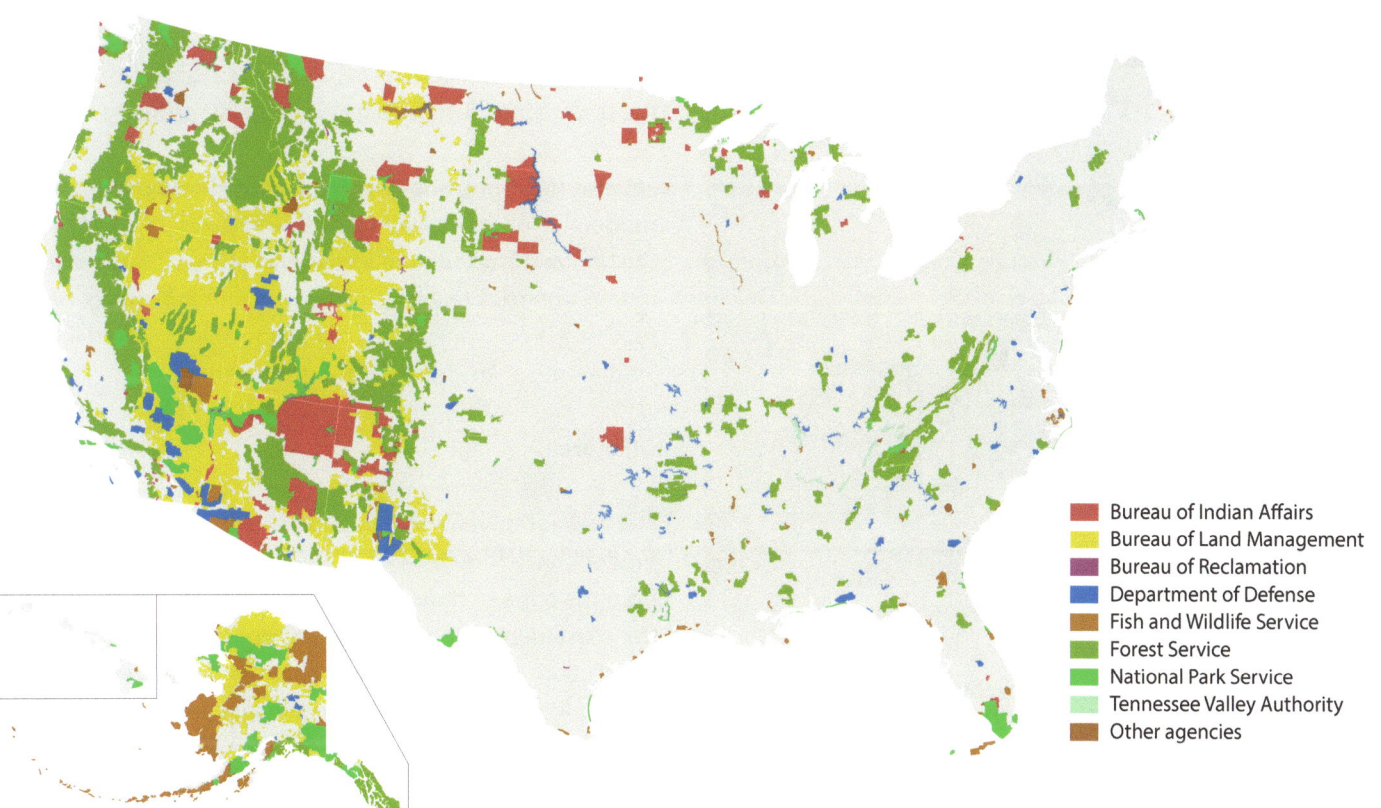

Federal and Native American lands in the United States. Colors indicate which federal agency oversees and regulates activities on these lands. Image credit: U.S. Geological Survey.[13]

Petroleum and the Environment
Part 21: U.S. Regulation of Oil and Gas Operations

designed to prevent the emission of a group of chemicals called volatile organic compounds (VOC),[15] and also capture methane, a potent greenhouse gas. Additional restrictions on methane and VOC emissions in the oil and gas industry that were issued in 2016 are delayed in legal disputes as of early 2018.[16]

- The EPA's Underground Injection Control program authorizes most states to regulate wells that dispose of oilfield waste, including produced water and hydraulic fracturing fluids that flow back up the well. However, the EPA itself regulates these wells in four oil- and gas-producing states: Pennsylvania, Virginia, New York, and Michigan. The Underground Injection Control program aims to protect groundwater from contamination. It does not address earthquakes caused by underground wastewater injection on non-federal lands.[18]

On Federal Land (Onshore)

The Bureau of Land Management (BLM) has jurisdiction over almost all leasing, exploration, development, and production of oil and gas on federal and Native American lands. BLM rules and standards for drilling and production[19] require all operations on federal land to comply with state and local regulations and protect life, property, and environmental quality. As of 2018, some federal drilling and production regulations enacted, revised, or proposed since 2008 are being re-evaluated or rescinded by the current Administration.[20]

The National Park Service regulates the small amount of oil and gas activity in National Parks (roughly 550 active wells in 2015[21]), where the federal government owns the land surface but not the underlying oil, natural gas, or mineral resources.[22]

Federal decisions about specific constraints on drilling and production on federal land (onshore and offshore) are based on the National Environmental Policy Act (1970).[23] This act requires federal agencies to assess the environmental impact of major federal actions, mainly by producing Environmental Impact Statements or Environmental Assessments.

Offshore

The federal government regulates offshore exploration and production for the Outer Continental Shelf (OCS), which extends from the edge of state waters (either 3 or 9 nautical miles from the coast, depending on the state) out to the edge of national jurisdiction, 200 nautical miles from shore.[24] The Bureau of Ocean Energy Management (BOEM) manages federal OCS leasing programs, conducts resource assessments, and licenses seismic surveys.[25] The Bureau of Safety and Environmental Enforcement (BSEE) regulates all OCS oil and gas drilling and production. These two agencies, plus the Office of Natural Resources

An advanced offshore blowout preventer (BOP). A BOP is a large, heavy set of valves fitted at the top of the well; if high pressures in the well overcome all other barriers, the BOP is designed to close off the well and the drill pipe to prevent oil or gas from escaping. BOPs of some kind are used on all oil and gas wells. In offshore drilling, the BOP is set on the seafloor or below the drilling-rig deck. Onshore, the BOP is connected to the top of the wellbore (below the drilling-rig deck). Image credit: Bureau of Safety and Environmental Enforcement.[17]

Petroleum and the Environment
Part 21: U.S. Regulation of Oil and Gas Operations

Revenue, which collects and disburses rents and royalties from offshore and onshore federal and Native American lands, were formed in the 2010 and 2012 reorganizations of the Minerals Management Service.

BSEE drilling and production regulations have been extensively revised in response to the 2010 Deepwater Horizon blowout and oil spill and a National Academies assessment of ways to prevent such incidents in the future.[26] The regulations include requirements for enhanced well design, improved blowout preventer design, testing and maintenance, and an increased number of trained inspectors.[27] The current Administration is in the process of reviewing and revising these regulations.

Regulation of Transportation, Storage, Refining, and Marketing

States regulate the operation of oil pipelines, as well as the construction and operation of natural gas gathering lines (small pipelines that move gas from the well to a processing facility or transmission line).[28,29] The federal Department of Transportation (DOT)'s Pipeline and Hazardous Materials Safety Administration regulates the operation of natural gas pipelines that provide long-distance transmission and local customer distribution,[30] as well as underground natural gas storage.[31] The EPA regulates air emissions from refineries and fuel distribution systems, including pipelines, trucks, and fuel dispensing facilities or service stations.[32]

The Federal Energy Regulatory Commission (FERC)[33] regulates the transportation of oil through interstate oil pipelines but does not oversee pipeline operations. FERC also reviews applications for the construction and operation of natural gas pipelines and liquefied natural gas (LNG) export and import terminals to certify their compliance with safety and environmental laws.

The Federal Railroad Administration (part of the DOT) is responsible for railroad safety, including rail transport of crude oil and refined products. The Pipeline and Hazardous Materials Safety Administration (DOT) and the Transportation Security Administration (Department of Homeland Security) issue safety standards for railroads. Approximately 11% of crude oil and petroleum products were transported by rail in 2014, up from 2.6% in 2009.[34]

Laws and Regulations Setting Energy Preferences

States and the federal government set requirements to encourage the use of particular energy types. Twenty-nine states have renewable portfolio standards that require electric utilities to sell a minimum percentage or amount of renewable energy.[35] On the federal level, the Renewable Fuels Program is overseen by the EPA in consultation with the Department of Agriculture. The Energy Policy Act of 2005 and the Energy Independence and Security Act of 2007 require that set amounts of renewable fuel be used in place of gasoline, heating oil, or jet fuel.[36] The applicable renewable fuels include ethanol, cellulosic biofuel, diesel from biomass, and other advanced biofuels. In 2018, the standards set by the EPA under authorization of the 2005 and 2007 acts require the use of 19.29 billion gallons of renewable fuels.[37] This results in most gasoline containing 10% ethanol (E10). E15 and E85, with 15% or up to 85% ethanol, respectively, are locally available and can be used in some vehicles.[38]

Other Regulations

Regulations not covered here include worker safety rules set by the Occupational Safety and Health Administration (see "Health and Safety in Oil and Gas Extraction" in this series for more information); regulations protecting antiquities and historic and religious sites; and the Endangered Species Act.[39] In addition, the oil and gas industry is bound by state and federal financial reporting laws and tax regulations.[40] Because oil and natural gas are traded globally and many oil and gas companies operate internationally, international trade rules and the laws of various other countries also impact the U.S. oil and gas industry.

References & More Resources

For a complete listing of references, see the "References" section of the full publication, *Petroleum and the Environment*, or visit the online version at: www.americangeosciences.org/critical-issues/petroleum-environment

Joy, M.P. and Dimitroff, S.D. (2016). Oil and gas regulation in the United States: overview. Westlaw, June 1, 2016.
https://content.next.westlaw.com/Document/I466099551c9011e38578f7ccc38dcbee/View/FullText.html

Petroleum and the Environment
Part 22

Health and Safety in Oil and Gas Extraction
Reducing the exposure of oil and gas workers to health and safety hazards

Introduction
Hundreds of thousands of people work in oil and gas extraction in the United States;[1] ensuring their health and safety is a major concern for employers, regulators, trade associations, industry groups, and local communities. Work in this industry involves physical labor, 24/7 operations, heavy machinery, hazardous chemicals, often-remote locations, and all weather conditions, resulting in an elevated risk of physical harm and the need for special protections to reduce this risk.

Physical Safety: Fatalities
From 2007 to 2016, more than 1,000 workers were killed in oil and gas extraction operations, a fatality rate six times higher than the average rate for all U.S. workers (21.6 vs. 3.5 per 100,000 workers).[2] Transportation events were the leading cause of death during this time period, making up 42% of all fatalities; most of these were the result of motor vehicle crashes. Worker fatalities also resulted from contact with objects/equipment (25%), fires/explosions (14%), exposure to harmful substances/environments (9%), and falls (8%). While the industry's fatality rate remains high, it decreased by 36% between 2003 and 2013 – a period during which the industry workforce was growing rapidly – suggesting that safety efforts may be yielding positive results.[3]

Chemical Exposure: Health Hazards
While fatal work injuries have been well studied, less is known about other health hazards. Since 2010, the National Institute for Occupational Safety and Health (NIOSH) has conducted field studies in partnership with industry to better identify chemical exposure hazards.[4] The major hazards identified through these studies were respirable crystalline silica dust during hydraulic fracturing and exposure to hydrocarbon gases and vapors when manually sampling oilfield tanks:[5]

Silica dust – Large quantities of silica sand are used during hydraulic fracturing. Loading and transferring this sand at the well site generates respirable-sized silica dust particles in concentrations that may exceed occupational exposure limits.[6]

Silica dust clouds from delivery trucks loading into sand movers at a hydraulic fracturing site. Inhalation of this silica is a major hazard associated with oil and gas operations. Image credit: Michael Breitenstein, National Institute for Occupational Safety and Health.[7]

Other processes that generate silica dust at the well site may include drilling with air and mixing cement to construct or plug a well. Inhalation of silica dust is associated with silicosis, other respiratory issues, and potentially other adverse health effects.[8] NIOSH recommends monitoring worker exposure and, when necessary, controlling exposure with engineering controls and improvements to work practices and procedures.[9] New Occupational Safety and Health Administration (OSHA) standards for silica take effect on June 23, 2021 for hydraulic fracturing operations, which are expected to implement engineering solutions that limit silica exposure.[10]

Hydrocarbon gases and vapors – Tanks holding crude oil or produced water are common in the oilfield. These tanks may be manually measured and sampled, which may expose workers to dangerous levels of hydrocarbon gases and vapors given off by these liquids. Between 2010 and 2014, at least nine oilfield workers died as a result of this exposure. In response, NIOSH and OSHA published a hazard alert related to manual tank gauging, which recommends using alternative systems to measure and sample

Petroleum and the Environment
Part 22: Health and Safety in Oil and Gas Extraction

Worker gauging an oilfield tank. Tank sampling is one of the main ways in which oilfield workers may be exposed to harmful vapors. Image credit: Photo courtesy CDPH and NIOSH.[12]

tank fluids without opening the tank hatch, as well as training workers and not permitting employees to inspect tanks alone.[11]

Other hazards include **hydrogen sulfide** gas (which occurs naturally in oil and natural gas and is extremely hazardous when inhaled); **noise** (from heavy machinery, for which OSHA sets maximum limits and required hearing protection[13]); and **diesel exhaust** (from drilling rigs and other equipment – while diesel exhaust is not specifically regulated, OSHA sets exposure limits for many of the most harmful air pollutants found in diesel exhaust[14]).

Long-term Health Hazards

Oilfield fluids contain a wide range of hazardous chemicals. While some can have immediate health effects (such as hydrogen sulfide gas, which can kill instantly at high concentrations), others may have longer-term effects (such as benzene, which is carcinogenic[15]). However, few published studies exist that track the long-term health consequences of working in oil and gas extraction, making it difficult to draw conclusions about specific long-term health risks.

Ongoing Mitigation Efforts

OSHA sets and enforces workplace standards for all industries, including oil and gas extraction. All operations are required to obtain or provide Safety Data Sheets for all hazardous chemicals and materials,[16] describing the associated risks and providing guidance and recommended procedures for minimizing exposure and addressing accidents.

In addition to the work from OSHA and NIOSH described above, the size and inherent risks of the oil and gas extraction industry mean there are a variety of ongoing efforts to research health and safety risks and implement solutions:

- Co-ordinated by NIOSH, the National Occupational Research Agenda (NORA) Oil and Gas Extraction Sector Council brings together representatives from industry, trade associations, academia, and major insurance companies to guide and conduct research on health and safety in the industry.[17]
- Trade associations such as the American Petroleum Institute (API),[18] the Association of Energy Service Companies,[19] and the International Association of Drilling Contractors[20] have programs that provide and update health and safety guidelines, convene meetings to share developments, and promote innovation and improvement in working conditions across the industry.
- The National Service, Transmission, Exploration & Production Safety (STEPS) network is a volunteer-run organization that brings together operators, contractors, trade associations, and educators to share best practices, discuss incidents, improve communications, and develop projects to work on specific issues.[21]

References & More Resources

For a complete listing of references, see the "References" section of the full publication, *Petroleum and the Environment*, or visit the online version at: www.americangeosciences.org/critical-issues/petroleum-environment

Mason, K.L. et al. (2015). Occupational Fatalities During the Oil and Gas Boom – United States, 2003-2013. Centers for Disease Control and Prevention, Morbidity and Mortality Weekly Report, 64(2), 551-554. https://www.cdc.gov/Mmwr/preview/mmwrhtml/mm6420a4.htm

National Institute for Occupational Safety and Health – Oil and Gas Extraction Program. https://www.cdc.gov/niosh/programs/oilgas/default.html

Petroleum and the Environment
Part 23

Subsurface Data in the Oil and Gas Industry
Probing beneath the Earth's surface for exploration and hazard mitigation

Introduction

Drilling for oil and gas is expensive. A single well generally costs $5-8 million onshore and $100-200 million or more in deep water.[1] To maximize the chances of drilling a productive well, oil and gas companies collect and study large amounts of information about the Earth's subsurface both before and during drilling. Data are collected at a variety of scales, from regional (tens to hundreds of miles) to microscopic (such as tiny grains and cracks in the rocks being drilled). This information, much of which will have been acquired in earlier exploration efforts and preserved in public or private repositories, helps companies to find and produce more oil and gas and avoid drilling unproductive wells, but can also help to identify potential hazards such as earthquake-prone zones or areas of potential land subsidence and sinkhole formation.

Mapping the Subsurface 1: Regional Data from Geophysics

In the 21st century, much is already known about the distribution of rocks on Earth. When looking for new resources, oil and gas producers will use existing maps and subsurface data to identify an area for more detailed exploration. A number of geophysical techniques are then used to obtain more information about what lies beneath the surface. These methods include measurements of variations in the Earth's gravity and magnetic field, but the most common technique is seismic imaging.

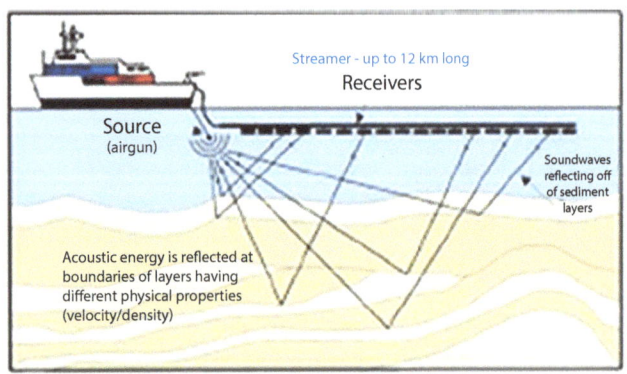

A typical setup for offshore seismic imaging. Image credit: U.S. Bureau of Ocean Energy Management.[2]

Seismic images are like an ultrasound for the Earth, and provide detailed regional information about the structure of the subsurface, including buried faults, folds, salt domes, and the size, shape, and orientation of rock layers. They are collected by using truck-mounted vibrators or dynamite (onshore), or air guns towed by ships (offshore), to generate sound waves; these waves travel into the Earth and are reflected by underground rock layers; instruments at the surface record these reflected waves; and the recorded waves are mathematically processed to produce 2-D or 3-D images of subsurface features. These images, which cover many square miles and have a resolution of tens to hundreds of feet, help to pinpoint the areas most likely to contain oil and/or gas.

Mapping the Subsurface 2: Local Data from Well Logs, Samples, and Cores

Drilling a small number of exploratory holes or using data from previously drilled wells (common in areas of existing oil and gas production) allows geologists to develop a much more complete map of the subsurface using well logs and cores.

A **well log** is produced by lowering geophysical devices into a wellbore, before (and sometimes after) the steel well casing is inserted, to record the rock's response to electrical currents and sound waves and measure the radioactive and electromagnetic properties of the rocks and their contained fluids.[3] Well logs have been used for almost 100 years[4] and are recorded in essentially all modern wells.

A **core** is a cylindrical column of rock, commonly 3-4 inches in diameter, that is cut and extracted as a well is drilled. A core provides a small cross-section of the sequence of rocks being drilled through, providing more comprehensive information than the measurements made by tools inside the wellbore.[5] Core analysis gives the most detailed information about the rock layers, faults and fractures, rock and fluid compositions, and how easily fluids (especially oil and gas) can flow through the rock and thus into the well.

Petroleum and the Environment
Part 23: Subsurface Data in the Oil and Gas Industry

Figure Caption: A box containing 9 feet of 4-inch diameter core from the National Petroleum Reserve, Alaska, showing the fine-scale structure and composition of the rock layers being drilled. Image credit: U.S. Geological Survey.[6]

By comparing the depth, thickness, and composition of subsurface rock formations in nearby wells, geoscientists can predict the location and productive potential of oil and gas deposits before drilling a new well. As a new well is being drilled, well logs and cores also help geoscientists and petroleum engineers to predict whether the rocks can produce enough oil or natural gas to justify the cost of preparing the well for production.[7]

Data Preservation

Preservation of subsurface data is an ongoing challenge, both because there is so much of it and because a lot of older data predate computer storage. A modern seismic survey produces a few to thousands of terabytes of data;[8] state and federal repositories collectively hold hundreds of miles of core;[9] and millions of digital and paper records are housed at state geological surveys. For example, the Kansas Geological Society library maintains over 2.5 million digitized well logs and associated records for the state.[10] Oil companies also retain huge stores of their own data. Preserving these data, which cost many millions of dollars to collect, allows them to be used in the future for a variety of purposes, some of which may not have been anticipated when the data were originally collected. For example, the shale formations that are now yielding large volumes of oil and natural gas in the United States were known but not considered for development for decades while conventional oil and gas resources were being extracted in many of the same areas. Archived well logs from these areas have helped many oil and gas producers to focus in on these shale resources now that the combination of hydraulic fracturing and horizontal drilling allow for their development.

Data for Hazard Mitigation

Oil and gas exploration is a major source of information about the subsurface that can be used to help identify geologic hazards:

- Since 2013, the oil and gas industry has provided more than 2,500 square miles of seismic data to Louisiana universities to assist with research into the causes and effects of subsidence in coastal wetlands. For example, seismic and well data have been used to link faults to historic subsidence and wetland loss near Lake Boudreaux.[11]

- To improve earthquake risk assessment and mitigation in metropolitan Los Angeles, scientists have used seismic and well data from the oil and gas industry to map out previously unidentified faults. This work was motivated by the 1994 Northridge earthquake, which occurred on an unknown fault that was not visible at the Earth's surface.[12]

References & More Resources

For a complete listing of references, see the "References" section of the full publication, *Petroleum and the Environment*, or visit the online version at: www.americangeosciences.org/critical-issues/petroleum-environment

U.S. Geological Survey – National Geological and Geophysical Data Preservation Program. https://datapreservation.usgs.gov/

Petroleum and the Environment
Part 24

Geoscientists in Petroleum and Environment
Geoscience informs all aspects of petroleum production and environmental protection

Introduction
Geoscience – the study of the Earth – underpins our understanding of the many intersections between petroleum and the environment, from the search for resources to the study of air pollutants. Without the work of geoscientists, we would have neither the energy system nor the environmental protections we benefit from today.

The Roles of Geoscientists
Finding oil and gas – petroleum geoscientists (**geophysicists**, **geologists**, and **geochemists**) work in multidisciplinary teams to decide where to perform seismic imaging (like an ultrasound of the Earth), collect and analyze seismic data, and analyze pre-existing drillhole data from wells to develop a detailed picture of the oil- or gas-rich rocks deep beneath the Earth's surface. These geoscientists use their knowledge of **stratigraphy** and **sedimentary processes** to predict the location and structure of oil- and gas-bearing rocks. **Structural geology** is used to predict the folding, faulting, and fracturing of rocks in order to interpret the shape of the oil- and gas-rich zones, identify areas where oil and gas may have migrated along faults and fractures, and improve the design of hydraulic fracturing operations. **Geochemistry** is used to study rock samples and fluids to better understand of the types and amounts of oil and gas present in the rocks. **Paleontologists** study fossils of ancient organisms, pollen grains, and more to help determine the age of rocks and how they formed. **Petroleum geologists** work closely with petroleum engineers who must ultimately design how the well will be drilled and prepared for production.

Drilling safely and effectively – drilling and preparing a well for production is generally managed by engineers. **Geologists** help to monitor the drilling by studying rock samples brought up to the surface during the drilling process and analyzing geophysical data obtained by instruments inside the well. **Hydrogeologists** analyze the risk of ground- and surface water contamination during the drilling and operation of the well, informing decisions about where to position a well and how to reduce the risk from surface and drilling operations.

Understanding and optimizing hydraulic fracturing – **geophysicists** and engineers study how fractures form during hydraulic fracturing, making the process more effective at extracting resources and improving understanding of the potential risks of contaminating groundwater or triggering earthquakes. **Seismologists** and engineers study the fractures and tiny earthquakes (microseisms) generated by hydraulic fracturing to monitor the progress of the operation, identify potential risks, and improve future operations.

Monitoring and mitigating leaks, spills, emissions, and other hazards – **geochemists** study the groundwater and soil in areas of oil and gas activity to identify potential leaks and spills. **Hydrologists** provide insight into how a source of contamination may spread through ground- or surface water, and provide the same expertise to help mitigate the impacts of spills and guide cleanup operations. **Atmospheric scientists** monitor the emissions of methane, toxic gases, and other pollutants from oil and gas operations, transportation infrastructure, refineries, and end-users (power plants, cars, etc.). **Geotechnical engineers** help facilities engineers to design oilfield and transportation infrastructure to reduce surface impacts and decrease the chances of harmful leaks, spills, and other equipment failures due to soil movement, subsidence, earthquakes, landslides, flooding, and other geohazards. **Geophysicists** and **geologists** study the earthquakes triggered by oil and gas operations to determine the causes and effects and so help to prevent or reduce future earthquakes. Many state and federal regulators are geoscientists with the knowledge and experience to develop and enforce rules that are consistent with local and regional geologic conditions and protect the environment, human health, and property rights.

Determining environmental impacts – **geochemists**, **hydrologists**, **soil scientists**, **oceanographers**, **atmospheric scientists**, and public health researchers study and forecast the short- and long-term effects of environmental contamination to identify the relative hazards from different sources and inform decisions

Petroleum and the Environment
Part 24: Geoscientists in Petroleum and Environment

about monitoring, mitigation, and remediation. **Climate scientists** study the short- and long-term effects of carbon dioxide, methane, and other major emissions on Earth's climate, informing many research, policy, and commercial decisions that extend beyond energy and the environment to issues such as diplomatic relations, foreign aid, resilience to natural hazards, infrastructure development, food and water supplies, and much more.

Education and outreach – Our present-day and future understanding of energy resources and the environment depends on well-trained geoscientists. This training requires experienced educators and the opportunity for direct participation in research and operations in private companies, universities, and non-profit organizations. Geoscience educators and outreach specialists also engage with the wider public, from schoolchildren and communities to elected representatives and nations, providing the scientific background to help inform decisions about energy and the environment at all levels.

To provide a sample of the various roles and responsibilities of geoscientists in petroleum and the environment, the rest of this section features profiles of seven geoscientists working in petroleum and the environment across many disciplines, from seismology to atmospheric chemistry, and across the United States, from Ohio to Alaska.

Profile 1: Bridget Scanlon
Senior Research Scientist, Bureau of Economic Geology, Jackson School of Geosciences, University of Texas at Austin.

I have worked on many different water resource issues throughout my 30-year career but most recently I have been focusing on water and energy interdependencies. At the Bureau of Economic Geology (BEG), we emphasize that the overlap between academia, industry, and government and collaboration among these groups is extremely important in advancing our understanding of the water-energy nexus.

My work for the BEG on the water-energy nexus began through collaboration with Dr. J.P. Nicot who conducted the first study to quantify water use for hydraulic fracturing to extract natural gas

in the Barnett Shale Play. As the development of unconventional resources grew, some national and global studies indicated that water scarcity was a particular concern in unconventional oil and gas production. To thoroughly address this issue we evaluated water supplies for shale oil and gas development in the Bakken and Eagle Ford plays and showed that water management systems were expanding to meet demand. In the Bakken, infrastructure was being developed to transport water from Lake Sakakawea, the 3rd largest reservoir in the U.S.; in the Eagle Ford, large brackish groundwater supplies were being tapped.

Our most recent work has focused on the Permian Basin, which is transitioning from one of the largest conventional resources to the largest global unconventional resource. Water management here is challenging, both in sourcing water for hydraulic fracturing in this semiarid region and managing large volumes of water that are co-produced with oil and gas, termed "produced water". Traditionally produced water has been reinjected into conventional reservoirs for enhanced oil recovery; however, the low permeability of unconventional reservoirs mean that we need to inject produced water from these reservoirs into other, non-producing rock layers, increasing subsurface pressures and potentially inducing earthquakes. I have been collaborating with colleagues at the Oklahoma Geological Survey and Stanford University to understand the controls on induced seismicity in Oklahoma and to determine how we can apply lessons from Oklahoma to other tight oil plays to minimize potential seismicity.

Petroleum and the Environment
Part 24: Geoscientists in Petroleum and Environment

Lastly, we have been working on water consumption for electricity generation, showing that water used for shale gas extraction is a small fraction (~5%) of the amount of water consumed in power plants to generate electricity with natural gas. I thoroughly enjoy working in a collaborative research team at the Bureau of Economic Geology, where we have strong connections with industry and government agencies and opportunities to develop fundamental understanding of the water-energy nexus.

Profile 2: David Houseknecht
Senior Research Geologist, U.S. Geological Survey

How much oil lies undiscovered beneath the environmentally fragile landscape of the Alaskan Arctic? Will development of those oil resources pump new life into the Trans Alaska Pipeline System, increase the nation's energy independence, and revive Alaska's sputtering economy? Will increased industry activity hasten deterioration of permafrost beneath the coastal plain, accelerate shoreline erosion, and jeopardize the habitat of polar bears, caribou, and migratory birds? These are just a few of the questions surrounding one of the nation's most politically volatile earth-science issues.

I have worked as a research geologist since 1995 on, and led since 2008, the USGS "Alaska Petroleum Systems" project charged with answering the "how much oil" question. The project includes about a dozen geoscientists scattered among USGS offices in Alaska, California, Colorado, and Virginia. And, I have represented the U.S. Geological Survey (USGS) perspective on Arctic Alaska petroleum resource potential since 1992 – to the Administration, Congress, Federal and State agencies, non-government organizations, the petroleum industry, the media, and the public – as political winds have swirled in every possible direction.

Despite including North America's largest oil field (Prudhoe Bay) and numerous other fields that would be considered huge by lower-48 standards, Arctic Alaska is lightly explored outside a core area along the Arctic coast. Consequently, our project conducts fundamental research to reconstruct the stratigraphic and structural evolution of the region by integrating subsurface mapping and interpretation, helicopter-supported field work, and sophisticated laboratory analyses to build a three-dimensional

geological framework. Within that framework, our interdisciplinary team uses a "source to sink" approach to estimate undiscovered oil potential by reconstructing oil-generative "kitchens", interpreting petroleum-migration pathways from kitchens to potential traps, and deciphering how the entire petroleum system evolved through time.

Perhaps the most challenging part of my job is communicating complex science, including probabilistic estimates of undiscovered petroleum resources, in a politically volatile environment. In doing so, I follow a handful of benchmark principles. First, I insist that our project scientists publish peer-reviewed reports on their research, preferably in prestigious journals. Second, we strive to make the assessment process as transparent as possible by communicating the science and methods often and to a broad spectrum of stakeholders. Third, we follow the USGS assessment protocol, which includes public review meetings to solicit feedback from external experts, thorough internal review of geological concepts, and rigorous statistical analysis for generating results. Fourth, we follow USGS Fundamental Science Practices to assure scientific integrity in conducting assessments and releasing results. Fifth, we communicate the results broadly and make ourselves available to discuss the results with stakeholders. These five steps bolster the scientific credibility of USGS assessments, underscore that resource assessments are rooted in science rather than political or societal influences, and create opportunities for collaborative application of assessment results to a broad range of resource and environmental topics.

Petroleum and the Environment
Part 24: Geoscientists in Petroleum and Environment

vehicles so that environmental investigations could be completed more efficiently on site without waiting for long laboratory turnaround times to decide "where to drill next."

My experience and knowledge gained during my work on trace gas analysis for oil and gas exploration allowed me to interpret both natural and anthropogenic stray gas migration from coal bed methane fields, conventional oil and gas fields, and directionally drilled shale oil/gas wells. I have been able to apply this technology to provide forensic-type analysis of these issues. I have been able to apply and share this approach in many contexts, and my work in this area led to my nomination and appointment to serve on the US-EPA Science Advisory Board (SAB) on the "EPA's Study of the Potential Impacts of Hydraulic Fracturing on Drinking Water Resources." With my experience in both drilling oil and gas wells and forensic geochemical investigation of stray gas migration, I was able to bring a unique perspective to this panel of 35 experts.

Profile 3: John Vincent Fontana
Professional Geologist, President/CEO, Vista GeoScience

I started my career in 1981 as a mud logger (wellsite geologist) for a company right after graduation. I later transitioned into their soil gas exploration geochemistry division where I helped develop exploration methods and used geostatistical data to evaluate micro-seeps for oil, gas and mineral prospecting. In 1986, I started my own company, continuing with soil gas exploration services, but expanded the services into environmental site investigation and remediation, and eventually ended up as sole owner of an environmental site investigation and remediation company.

The soil gas exploration techniques I developed and used for exploration turned out to be applicable to environmental contaminant site investigations for petroleum and industrial chemical spills, gas migration at landfills, and natural gas migration issues. We went from using hand driven soil gas probes and augers to truck- or tractor-mounted drilling rigs for collecting soil gas, ground water and soil samples for environmental analysis using EPA-approved lab methods. To make investigations more efficient, we moved laboratory bench instruments (mainly gas chromatographs and portable spectrometers) into mobile

Profile 4: Katherine M. Saad
Postdoctoral Fellow, The Aerospace Corporation

A camera captures light; a hyperspectral image captures electromagnetic waves far beyond what we can see. In this way, hyperspectral images are often likened to photographs, but to me they are more like a secret code that chemicals in the atmosphere are sending to our sensors. These chemicals have unique signatures based on the ways they absorb and emit infrared light, and detecting their signals allows scientists to map emissions released from both natural processes and human activities on the Earth's surface. Our team at The Aerospace Corporation deploys our sensors on aircraft, enabling us to scan large areas and get a picture of emissions plumes from above. This makes our instruments well suited for measuring unintended emissions from oil and gas wells, processing facilities, and distribution infrastructure. By taking these measurements over large areas, researchers can map leaks and other unintended releases of hydrocarbons, which besides being potential safety hazards are also energy sources that are wasted when lost, and in some cases present the risk of exposure to toxic chemicals.

As a principal investigator (PI), I design research studies and am responsible for the data collection, calibration, and analysis involved. The most exciting of these are the field campaigns themselves – not just flying on the aircraft with the instrument, but also deciding on flight paths and thinking on my feet when challenges arise. In that way, my PhD in Environmental Science and Engineering prepared me far beyond providing a foundation for atmospheric chemistry research, as I learned how to think critically and act decisively in a fast-paced environment.

Prior to graduate school, I worked in the energy sector after earning a B.S. in Environmental Sciences, a B.A. in Political Science, and a minor in Energy and Resources. The core of the B.S. was physical science, but my program gave me the flexibility to take electives such as environmental impact assessment and remote sensing engineering. I also had the opportunity to pursue a thesis project in each of those three departments, which sparked my enthusiasm for field work.

Profile 5: Sherilyn Williams-Stroud
Structural Geologist, President & CEO, Confractus, Inc.

As a structural geologist in exploration and production, I specialized in fracture analysis. Rock fractures provide pathways for fluid flow into and out of a reservoir, so their presence/absence and the timing of their formation will determine whether or not a petroleum reservoir exists. If fracturing happened before a seal formed to trap the oil and/or gas, the hydrocarbons will simply pass through. However, if fracturing occurred after the reservoir formed, the natural fractures can make it much easier to extract oil or gas from that reservoir if the fracture network is

well understood. Understanding natural fracturing and its timing has also helped lead to a better understanding of some of the processes related to unconventional reservoirs. One issue I have focused on is trying to determine when a shale might become naturally fractured. As hydrocarbons form, they increase the internal pressure on the rock, which can create fractures in much the same way as human-caused hydraulic fracturing.

I joined a microseismic company at the beginning of the boom for unconventional oil and gas production, at a time when induced seismicity began to be seen as a potential hazard for surrounding communities. Although the vast majority of induced events from hydraulic fracturing were too small to be felt, some of the early hydraulic fracture mapping data showed that seismicity was occurring at a distance from the reservoir that could not have been related to the increased fluid pressure near the wellbore. We interpreted these events as fault plane reactivations from the stress change in the reservoir during and after the frack. In the early days, even those larger events were not large enough to be felt, but they raised the issue of how large an event could happen and how far away it could be. Because the boom in unconventionals resulted in a large increase in the amount of produced water in need of disposal, high rates of disposal by underground injection led to the surge in felt earthquakes in high-production regions like Texas and Oklahoma. Geomechanical and structural analysis, combined with microseismic monitoring and geologic interpretation of the data, has led to a better understanding of the existing fracture and fault networks involved in induced earthquakes.

Petroleum and the Environment
Part 24: Geoscientists in Petroleum and Environment

Microseismic monitoring is also key for the geothermal energy industry, improving our understanding of permeable pathway locations in engineered/enhanced geothermal systems (EGS) and the role of existing faults in induced seismicity. Recent developments in low-temperature geothermal resources have strengthened the transferability of microseismic monitoring between geothermal and oil/gas systems, as both low-temperature geothermal resources and petroleum are found in sedimentary basins.

Profile 6: Steven Dade
Geologist II, Ohio Department of Natural Resources: Division of Oil & Gas Resources Management

As a Geologist II for the Ohio Department of Natural Resources (ODNR): Division of Oil & Gas Resources Management (DOGRM), I've worked extensively in expanding, maintaining and operating the OhioNET Seismic Network. This network aims to tackle several major issues on induced seismicity hazards and environmental protection in Ohio. Since beginning as an intern in 2013 with ODNR, my role within the program has changed dramatically. As an intern, much of my early work focused on georeferencing old satellite imagery and learning how to construct seismic vaults and stations. Now as a stand-alone program within the division, OhioNET has expanded in both personnel and station coverage as oil and gas operations continue to grow in Ohio. This unique role has led to learning and understanding the interactions of a wide range of disciplines including seismology, geology, seismic station construction, seismic processing software, GIS mapping, telemetry, data acquisition, and data management.

The OhioNET seismic network began operation in 2012 and has become a statewide, real-time network consisting of over 45 seismic stations from multiple sources including ODNR, industry operators, Transportable Array (TA), and the Central Eastern United States Network (CEUSN).

To cope with and prepare for the increasing need for regulatory oversight in oil and natural gas operations related to the Ohio Utica shale play, much of the work done at OhioNET focuses on guaranteeing the ability to accurately locate and quantify seismic events in both disposal areas and unconventional oil- and gas-producing zones. Another key factor for me has been to stay up to date with industry techniques and technologies to help guide further expansion and effectiveness of this evolving field of study.

In terms of regulation, ODNR has strengthened its capacity through additional human resources, improved infrastructure, and a robust regulatory framework with improved policies. This leap in growth has improved our ability to perform our principal regulatory role for oil and gas operations in Ohio, as well as our contribution to the state's overall seismic risk reduction. These improvements and proactive steps will hopefully continue to build a culture of safety surrounding oil and gas extraction and further public and industry participation.

Profile 7: Susan Nissen
Consulting Geophysicist

I work as a consulting geophysicist for several small independent petroleum companies focusing on exploration and production in the midcontinent, primarily Kansas, Colorado, and Nebraska. I first became interested in pursuing a career in geophysics when I took an introductory geology class as a freshman in college. After receiving a B.S. in geophysics from the University of Delaware and a Ph.D. in marine geophysics from Columbia University, I spent a number of years in the petroleum industry as a research scientist at an industry research center in Tulsa, Oklahoma. While there, I was involved in the development and testing of new seismic interpretation techniques. In 1999, I moved to Kansas to work in the Petroleum Research Section of the Kansas Geological Survey, focusing my research on the application of new seismic techniques to reservoirs in the US midcontinent. I started my own consulting business in 2006.

Geophysicists use various indirect methods to study the rocks beneath the earth's surface. These methods complement the more direct, but spatially limited, information about the subsurface that is obtained by drilling a well. The geophysical method that I use is 3-D reflection seismology, or "3-D seismic". In reflection seismology, vibrations (seismic waves) are generated at the earth's surface using a source such as a vibrator truck. The seismic waves penetrate the subsurface and bounce off boundaries (horizons) between rock layers. The reflected seismic waves that return to the surface are then recorded by seismic receivers. The travel time (time from when the seismic wave is generated until it is recorded) and amplitude of each of these seismic reflections can be used to infer information about the subsurface. In 3-D seismic, the sources and receivers are set up to produce a closely-spaced spatial grid of data. I interpret the seismic reflections from key subsurface horizons, such as the tops of petroleum-producing formations, and make maps, similar to topographic maps, of the shapes of these horizons. I also integrate the seismic data with well data to predict the structure and physical properties (e.g., porosity, rock type) of petroleum reservoirs away from existing well locations. This information allows my clients to better position new drilling locations.

Note: all images provided courtesy of the featured geoscientists

More Resources

For more information on the relationships between petroleum and the environment, see the full publication, *Petroleum and the Environment*, in print or online at: www.americangeosciences.org/critical-issues/petroleum-environment

American Geosciences Institute – Geoscience Workforce Program. https://www.americangeosciences.org/workforce

Petroleum and the Environment
Glossary of Terms

Glossary of Terms

This section contains short explanations of some of the more technical terms used in this publication. For additional assistance with specific terms relating to petroleum and the environment, please refer to the following glossaries:

- U.S. Geological Survey - Energy Glossary and Acronym List. https://energy.usgs.gov/GeneralInfo/HelpfulResources/EnergyGlossary.aspx

- U.S. Energy Information Administration – Glossary. https://www.eia.gov/tools/glossary/

- Schlumberger – Oilfield Glossary. http://www.glossary.oilfield.slb.com/

- U.S. Environmental Protection Agency – Terms & Acronyms. https://iaspub.epa.gov/sor_internet/registry/termreg/searchandretrieve/termsandacronyms/search.do

Aquifer – subsurface gravel, sand, or rock that contains or transmits groundwater. This water may be fresh, brackish, or saline, depending on the aquifer.

Barrel – one barrel of oil equals 42 gallons. Roughly the volume of a small bathtub. Oil is no longer stored or transported in 42-gallon barrels, which were originally constructed from wood. The steel drums shown in many pictures hold 55 gallons.

Brackish water – slightly salty water containing 1 to 10 grams of dissolved solids per liter of water. Considered not fit for human consumption.

Cased/casing – steel pipe cemented inside a wellbore. Casing prevents the wall of the well from caving in; prevents inflow of unwanted fluids from surrounding rocks at different points in the well; and prevents oil, gas, and hydraulic fracturing fluid from leaking out of the well into surrounding rocks. Wells may have several nested layers ("strings") of casing extending to different depths for the various different purposes mentioned above.

Conventional – refers to oil and gas reservoirs in which wells can be drilled so that oil or gas flows naturally or can be pumped to the surface. The term reflects that this approach has been in common use since the late 19th century. Conventional oil and gas has the same range of compositions as unconventional oil and gas.

Drilling pad – a temporary drilling site where the drill rig and associated equipment operates. The pad is a cleared, flat area paved with gravel (or occasionally crushed shells, wood, asphalt or cement). After the drilling is completed, this material is removed from much of the site.

Feedstock – any raw material fed into a process for conversion into something different. For example, crude oil is a feedstock for a refinery producing gasoline, and petrochemicals are feedstocks for producing plastics.

Flowback water – water that flows back up the wellbore after a hydraulic fracturing operation. It consists of hydraulic fracturing fluid and produced water from the rocks being fractured.

Gas condensate – the composition of natural gas varies widely from place to place. Some natural gas contains components that are gases in the reservoir and when initially produced but liquids at lower pressures or temperatures (e.g., when they are cooled at the well site specifically to extract these valuable liquids). These "gas condensates" include propane, butane, pentane and hexane.

Green completions – using equipment to capture the gas and gas condensate that flows out of a well during completion operations. Captured gas and condensate can then be sold instead of being flared or vented to the atmosphere.

Petroleum and the Environment
Glossary of Terms

Hydraulic fracturing – a technique used to increase the flow of oil and/or gas out of rocks by creating additional fractures for oil and/or gas to flow through and into a wellbore. This is achieved by injecting fluid into the wellbore at very high pressures, causing the surrounding rocks to crack, which increases the number and size of flow paths into the wellbore. The injected fluid is typically water with added chemicals to improve performance and grains (commonly sand) that hold open the newly created fractures. Hydraulic fracturing has been performed since the 1940s in vertical and then slanted wells, but today is most commonly associated with horizontal drilling of shales and other low-permeability rocks.

Lease – In the oil and gas industry, leases are provided by government or private owners of mineral rights (land ownership may or may not include ownership of underlying resources) to companies, allowing them to explore for and produce oil and/or natural gas under certain conditions.

LIDAR – a remote sensing method (Light Detection and Ranging) that uses light from a pulsed laser to measure distances from the measuring system to an object or surface. In geoscience applications, LIDAR is commonly used to obtain precise information about land elevation or seafloor features.

Magnitude (earthquake) – a measure of the energy released at the source of an earthquake as determined from seismographic measurements. In contrast, *intensity* is the strength of shaking at a particular location. Higher-magnitude earthquakes produce more intense shaking, but intensity also depends on other factors, including the distance from the source of an earthquake and the type of rock being shaken. Magnitude 3 earthquakes may be widely felt. Magnitude 5 to 7 earthquakes may cause slight to major building damage. The largest earthquakes recorded are over magnitude 9 (e.g. the 2004 Sumatra earthquake and the 2011 Tohoku earthquake).

Natural gas liquids – Chemical compounds found in natural gas that are heavier than methane. The most common natural gas liquids (ethane and propane) are only liquid under high pressure (underground or when pressurized in tanks) and become gases at atmospheric pressure. Other substances commonly grouped as NGLs include naphtha, natural gasoline, and gas condensates.

Organic compound – a large class of chemicals in which one or more carbon atoms are linked to other elements, most commonly hydrogen, oxygen or nitrogen. *Hydrocarbons* – a subset of organic compounds consisting only of carbon and hydrogen – make up the majority of oil and gas by volume.

Orphaned well – a well that does not have any legally responsible or financially able party to deal with its plugging, abandonment, and reclamation responsibilities.

Permeability – the extent to which a material allows fluids to pass through it.

Play – a group of related oil fields in the same area.

Produced water – water that is produced as a byproduct along with oil and/or gas. Many oil or gas reservoirs contain water either in the oil or gas reservoir or in a zone underlying the reservoir. The water composition varies regionally, from almost fresh to many times saltier than seawater.

Refined petroleum products – products derived by refining crude oil. They include gasoline, kerosene, jet fuel, heating oil, fuel oil, petrochemical feedstocks, and lubricating oil.

Reserves – the quantities of oil or natural gas that are anticipated to be commercially recoverable from known sources. In discussions of availability, "reserves" are different from "resources," which are *estimated* to be *potentially* recoverable.

Reservoir – in the oil and gas industry, "reservoir" generally refers to a localized accumulation of rocks that are rich in oil and/or gas.

Saline water – water containing a high concentration of dissolved salts. Saline water is saltier than brackish water, containing greater than 1% (or 10 grams per liter) of dissolve salts, and is considered non-potable unless treated (desalinated) to remove the salts. Seawater has a salinity of 3.5%; saline groundwater may have significantly higher salt contents.

Petroleum and the Environment
Glossary of Terms

Shale – a fine-grained sedimentary rock that forms from the compaction of mud. Shale is made up of many thin layers. Shales containing organic material are commonly the source of oil and natural gas.

Subsidence – the gradual or sudden settling or sinking of the Earth's surface. Subsidence is often caused by the removal of water, oil, gas, coal, or mineral resources from the ground. Subsidence may also be caused by natural processes such as earthquakes, sediment compaction, or movement of surface landmasses along faults.

Subsurface – anything below the surface of the Earth. Synonyms: underground, subterranean.

Technically recoverable – that portion of the oil or natural gas in a reservoir that can be extracted using currently available technology and industry practices. Technically recoverable oil/gas may or may not be economic to produce (i.e., economically recoverable).

Unconventional – a loosely defined term referring to oil and gas extracted from difficult-to-produce reservoirs using newer, often complex or expensive techniques. In the U.S., most unconventional oil and gas is produced from low-permeability or "tight" shale or sandstone formations using horizontal wells and hydraulic fracturing.

Wellbore – a hole drilled into the Earth to aid in the exploration and recovery of natural resources including gas, oil, or water.

Well completion – the process of making a well ready for production. This commonly includes pumping clean fluid into the well to remove the drilling mud or residue from other treatments, perforating the well casing next to the oil-/gas-producing zone, injecting acids to remove drilling-mud residue or widen existing cracks, hydraulically fracturing the producing zone, and installing production tubing and pumps in the well.

Petroleum and the Environment References

References

Part 1: Petroleum and the Environment: an Introduction

1. U.S. Energy Information Administration – U.S. Energy Facts Explained. https://www.eia.gov/energyexplained/?page=us_energy_home
2. International Energy Agency (2017). Key world energy statistics, 2017. https://www.iea.org/publications/freepublications/publication/KeyWorld2017.pdf
3. U.S. Energy Information Administration (2018). Annual Energy Outlook 2018. https://www.eia.gov/outlooks/aeo/
4. U.S. Energy Information Administration (2012). Energy Perspectives 1949-2011. https://www.eia.gov/totalenergy/data/annual/perspectives.php
5. U.S. Energy Information Administration – Monthly Energy Review. https://www.eia.gov/totalenergy/data/monthly/index.php
6. U.S. Energy Information Administration – Natural Gas Gross Withdrawals and Production. https://www.eia.gov/dnav/ng/ng_prod_sum_a_EPG0_FGW_mmcf_m.htm
7. U.S. Energy Information Administration – Crude Oil Production. https://www.eia.gov/dnav/pet/pet_crd_crpdn_adc_mbblpd_a.htm
8. U.S. Pipeline and Hazardous Materials Safety Administration – Data and Statistics Overview. https://www.phmsa.dot.gov/data-and-statistics/pipeline/data-and-statistics-overview
9. U.S. Energy Information Administration – Petroleum & Other Liquids: Exports. https://www.eia.gov/dnav/pet/pet_move_exp_dc_NUS-Z00_mbblpd_m.htm
10. National Oceanic and Atmospheric Administration – Climate change impacts. http://www.noaa.gov/resource-collections/climate-change-impacts
11. NASA – The consequences of climate change. https://climate.nasa.gov/effects/
12. Intergovernmental Panel on Climate Change – Climate Change 2013: The Physical Science Basis. http://www.ipcc.ch/report/ar5/wg1/
13. U.S. Environmental Protection Agency – Global Greenhouse Gas Emissions Data. https://www.epa.gov/ghgemissions/global-greenhouse-gas-emissions-data
14. U.S. Environmental Protection Agency (2018). Inventory of U.S. Greenhouse Gas Emissions and Sinks: 1990-2016. https://www.epa.gov/ghgemissions/inventory-us-greenhouse-gas-emissions-and-sinks-1990-2016
15. U.S. Environmental Protection Agency – Assessment of the Potential Impacts of Hydraulic Fracturing for Oil and Gas on Drinking Water Resources. https://cfpub.epa.gov/ncea/hfstudy/recordisplay.cfm?deid=244651
16. National Research Council (2013). Induced Seismicity Potential in Energy Technologies. Washington, DC: The National Academies Press. https://www.nap.edu/catalog/13355/induced-seismicity-potential-in-energy-technologies
17. Foulger, G et al. (2017). Global review of human-induced earthquakes. Earth-Science Reviews, 178, 438-514. https://www.sciencedirect.com/science/article/pii/S001282521730003X
18. U.S. Geological Survey – Induced Earthquakes: Myths and Misconceptions. https://earthquake.usgs.gov/research/induced/myths.php
19. Atkinson, G.M. et al. (2016). Hydraulic Fracturing and Seismicity in the Western Canada Sedimentary Basin. Seismological Research Letters 87(3), 631-647. https://pubs.geoscienceworld.org/ssa/srl/article-abstract/87/3/631/315665/hydraulic-fracturing-and-seismicity-in-the-western
20. American Geosciences Institute (2017). State Responses to Induced Earthquakes. https://www.americangeosciences.org/policy-critical-issues/webinars/state-responses-induced-earthquakes
21. U.S. Energy Information Administration – What is U.S. electricity generation by energy source? https://www.eia.gov/tools/faqs/faq.php?id=427&t=3
22. "Steamflooding Keeps California Field Producing 117 Years Later." S. Weeden, E&P Magazine, April 1, 2016. https://www.epmag.com/steamflooding-keeps-california-field-producing-117-years-later-844191
23. U.S. Pipeline and Hazardous Materials Safety Administration – Incident Statistics. https://www.phmsa.dot.gov/hazmat-program-management-data-and-statistics/data-operations/incident-statistics

Part 2: Water in the Oil and Gas Industry

1. Maupin, M.A. et al. (2014). Estimated Water Use in the United States in 2010, U.S. Geological Survey Circular 1405. https://pubs.usgs.gov/circ/1405/
2. U.S. Geological Survey - How much water does the typical hydraulically fractured well require? https://www.usgs.gov/faqs/how-much-water-does-typical-hydraulically-fractured-well-require
3. Veil, J. (2015). U.S. Produced Water Volumes and Management Practices in 2012, http://www.veilenvironmental.com/publications/pw/prod_water_volume_2012.pdf
4. National Academies of Science, Engineering, and Medicine (2017). Flowback and Produced Waters: Opportunities and Challenges for Innovation: Proceedings of a Workshop. Washington DC: The National Academies Press. https://www.nap.edu/catalog/24620/flowback-and-produced-waters-opportunities-and-challenges-for-innovation-proceedings
5. U.S. Geological Survey - Wastewater Disposal Facility in Colorado. https://www.usgs.gov/media/images/wastewater-disposal-facility-colorado

Petroleum and the Environment References

6. Scanlon, BR, et al. (2014). Comparison of Water Use for Hydraulic Fracturing for Unconventional Oil and Gas versus Conventional Oil. Environ. Sci. Technol., 48, 12386-12393. http://pubs.acs.org/doi/abs/10.1021/es502506v

7. Clark, C.E. et al. (2013). Life Cycle Water Consumption for Shale Gas and Conventional Natural Gas. Environ. Sci. Technol., 47, 11829-11836. https://pubs.acs.org/doi/abs/10.1021/es4013855

Part 3: Induced Seismicity from Oil and Gas Operations

1. National Research Council (2013) - Induced Seismicity Potential in Energy Technologies. https://www.nap.edu/catalog/13355/induced-seismicity-potential-in-energy-technologies
2. Nicholson, C. and Wesson, R.L. (1990). Earthquake Hazard Associated with Deep Well Injection – A Report to the U.S. Environmental Protection Agency. U.S. Geological Survey Bulletin 1951. https://pubs.usgs.gov/bul/1951/report.pdf
3. American Geosciences Institute (2017). State Responses to Induced Seismicity. https://www.americangeosciences.org/policy-critical-issues/webinars/state-responses-induced-earthquakes
4. Atkinson, G. et al. (2016). Hydraulic Fracturing and Seismicity in the Western Canada Sedimentary Basin. Seismological Research Letters, 87(3), 631-647. https://pubs.geoscienceworld.org/ssa/srl/article-abstract/87/3/631/315665/hydraulic-fracturing-and-seismicity-in-the-western
5. U.S. Geological Survey – Induced Earthquakes: Overview. https://earthquake.usgs.gov/research/induced/overview.php
6. Veil, J. (2015). U.S. Produced Water Volumes and Management Practices in 2012. http://www.veilenvironmental.com/publications/pw/prod_water_volume_2012.pdf
7. U.S. Geological Survey – Do all wastewater disposal wells induce earthquakes? https://www.usgs.gov/faqs/do-all-wastewater-disposal-wells-induce-earthquakes
8. U.S. Geological Survey (2017) – Short-term Induced Seismicity Models. https://earthquake.usgs.gov/hazards/induced/
9. U.S. Geological Survey - Earthquake Catalog. https://earthquake.usgs.gov/earthquakes/search/
10. U.S. Geological Survey – House Damage from 2011 Oklahoma Earthquake. https://www.usgs.gov/media/images/house-damage-2011-oklahoma-earthquake
11. Kansas Geological Survey (2017). Update on Kansas Seismicity: A Year of Change, What Does it Mean? http://www.kgs.ku.edu/PRS/Seismicity/2017/KS_Seismicity_KDHE_Wichita_08_30_17_V4.pdf
12. Kansas Geological Survey (2018). Number of Earthquakes Each Month, 2013 to Date. http://www.kgs.ku.edu/PRS/Seismicity/2018/Kansas_earthquake_frequency.pdf
13. Frohlich, C. et al. (2016). A Historical Review of Induced Earthquakes in Texas. Seismological Research Letters, 87 (4), 1022-1038. https://pubs.geoscienceworld.org/ssa/srl/article-abstract/87/4/1022/314110/a-historical-review-of-induced-earthquakes-in
14. Bao, X. and Eaton, D.W. (2016). Fault activation by hydraulic fracturing in western Canada. Science, 354, 1406-1409. http://science.sciencemag.org/content/early/2016/11/16/science.aag2583.full
15. Skoumal, R. et al. (2015). Earthquakes Induced by Hydraulic Fracturing in Poland Township, Ohio. Seismological Society of America Bulletin, 105(1), 189-197. https://pubs.geoscienceworld.org/ssa/bssa/article-abstract/105/1/189/323441
16. Ground Water Protection Council and Interstate Oil and Gas Compact Commission (2015). Potential Injection-Induced Seismicity Associated with Oil and Gas Development: A Primer on Technical and Regulatory Considerations Informing Risk Management and Mitigation. http://www.gwpc.org/sites/default/files/finalprimerweb.pdf
17. U.S. Environmental Protection Agency, Underground Injection Control National Technical Workgroup (2015). Minimizing and Managing Potential Impacts of Injection-Induced Seismicity from Class II Disposal Wells: Practical Approaches. https://www.epa.gov/sites/production/files/2015-08/documents/induced-seismicity-201502.pdf
18. Chen, X. et al. (2017). The Pawnee earthquake as a result of the interplay among injection, faults, and foreshocks. Scientific Reports, 7, 4945. https://www.nature.com/articles/s41598-017-04992-z

Part 4: Water Sources for Hydraulic Fracturing

1. U.S. Geological Survey Water Availability and Use Science Program, National Brackish Groundwater Assessment, What is Brackish? https://water.usgs.gov/ogw/gwrp/brackishgw/brackish.html
2. Maupin, M.A. et al. (2014). Estimated Water Use in the United States in 2010, U.S. Geological Survey Circular 1405. https://pubs.usgs.gov/circ/1405/
3. Union of Concerned Scientists (2011). Freshwater Use by U.S. Power Plants. Figure 9 used with permission. http://www.ucsusa.org/sites/default/files/legacy/assets/documents/clean_energy/ew3/ew3-freshwater-use-by-us-power-plants.pdf
4. U.S. Global Change Research Program (2014). U.S. National Climate Assessment: National Climate Change Impacts in the United States. http://nca2014.globalchange.gov/
5. Gallegos, T.J. et al. (2015). Hydraulic fracturing water use variability in the United States and potential environmental implications. Water Resources Research, 51(7), 5839-5845. https://agupubs.onlinelibrary.wiley.com/doi/full/10.1002/2015WR017278. Image obtained from https://www.usgs.gov/media/images/percentage-hydraulic-fracturing-use
6. Veil, J. (2015). U.S. Produced Water Volumes and Management Practices in 2012. http://www.veilenvironmental.com/publications/pw/prod_water_volume_2012.pdf
7. Energy & Environment Research Center, University of North Dakota (2016). A Review of Bakken Water Management Practices and Potential Outlook. https://www.undeerc.org/Bakken/pdfs/Bakken_Water_Management_Practices_and_Potential_Outlook.pdf
8. Whitfield, S. (2017). Permian, Bakken Operators Face Produced Water Challenges. Society of Petroleum Engineers. https://www.spe.org/en/print-article/?art=2982
9. Kondash, A.J. et al. (2017). Quantity of flowback and produced waters from unconventional oil and gas exploration. Science of The Total Environment, 574, 314-321. https://www.sciencedirect.com/science/article/pii/S004896971631988X

Petroleum and the Environment References

[10] Vidic, R.D. and Yoxtheimer, D. (2017). Changes in the Quantity and Quality of Produced Water from Appalachian Shale Energy Development and their Implications for Water Reuse. Presentation at the Pennsylvania State University Marcellus Center for Outreach and Research. http://www.shalenetwork.org/sites/default/files/Vidic_2017_ShaleNetwork.pdf

[11] National Academies of Science, Engineering, and Medicine (2017). Flowback and Produced Waters: Opportunities and Challenges for Innovation: Proceedings of a Workshop. Washington DC: The National Academies Press. https://www.nap.edu/catalog/24620/flowback-and-produced-waters-opportunities-and-challenges-for-innovation-proceedings

[12] Scanlon, B.R. et al. (2014). Will water scarcity in semiarid regions limit hydraulic fracturing of shale plays? Environmental Research Letters, 9, 124011. http://iopscience.iop.org/article/10.1088/1748-9326/9/12/124011

[13] FracFocus Chemical Disclosure Registry. http://fracfocus.org/

[14] Nicot, J-P. et al. (2011). Current and Projected Water Use in the Texas Mining and Oil and Gas industry. Bureau of Economic Geology for the Texas Water Development Board, Contract No. 0904830939. http://www.twdb.texas.gov/publications/reports/contracted_reports/doc/0904830939_MiningWaterUse.pdf

[15] Texas Water Development Board – Desalination Facts. http://www.twdb.texas.gov/innovativewater/desal/facts.asp

[16] American Geosciences Institute (2016). Desalination as a Source of Fresh Water. https://www.americangeosciences.org/policy-critical-issues/webinars/desalination-source-fresh-water

[17] Society of Petroleum Engineers PetroWiki - Fracturing fluids and additives. http://petrowiki.org/Fracturing_fluids_and_additives

Part 5: Using Produced Water

[1] U.S. Geological Survey – National Produced Waters Geochemical Database. https://energy.usgs.gov/EnvironmentalAspects/EnvironmentalAspectsofEnergyProductionandUse/ProducedWaters.aspx#3822349-data

[2] Veil, J. (2015). U.S. Produced Water Volumes and Management Practices in 2012. http://www.veilenvironmental.com/publications/pw/prod_water_volume_2012.pdf

[3] Oklahoma Water Resources Board – Water for 2060 Produced Water Working Group. https://www.owrb.ok.gov/2060/pwwg.php

[4] Colorado School of Mines Advanced Water Technology Center – Produced Water Beneficial Use Case Studies. http://aqwatec.mines.edu/produced_water/assessbu/case/

[5] National Academies of Science, Engineering, and Medicine (2017). Flowback and Produced Waters: Opportunities and Challenges for Innovation: Proceedings of a Workshop. Washington DC: The National Academies Press. https://www.nap.edu/catalog/24620/flowback-and-produced-waters-opportunities-and-challenges-for-innovation-proceedings

[6] Railroad Commission of Texas – Injection and Disposal Wells. http://www.rrc.state.tx.us/about-us/resource-center/faqs/oil-gas-faqs/faq-injection-and-disposal-wells/

[7] "PA DEP approved 11th underground injection well for oil and gas waste." J. Hurdle, NPR StateImpact, June 5, 2017. https://stateimpact.npr.org/pennsylvania/2017/06/05/pa-dep-approved-11th-underground-injection-well-for-oil-and-gas-waste/

[8] U.S. Environmental Protection Agency – Protecting Underground Sources of Drinking Water from Underground Injection (UIC). https://www.epa.gov/uic

[9] Vidic, R.D. and Yoxtheimer, D. (2017). Changes in the Quantity and Quality of Produced Water from Appalachian Shale Energy Development and their Implications for Water Reuse. Presentation at the Pennsylvania State University Marcellus Center for Outreach and Research. http://www.shalenetwork.org/sites/default/files/Vidic_2017_ShaleNetwork.pdf

[10] Colorado School of Mines Advanced Water Technology Center – Produced Water Beneficial Use Case Studies. http://aqwatec.mines.edu/produced_water/assessbu/case/

[11] National Research Council (2010). Management and Effects of Coalbed Methane Produced Water in the Western United States. Washington, DC: The National Academies Press. https://www.nap.edu/catalog/12915/management-and-effects-of-coalbed-methane-produced-water-in-the-western-united-states

[12] California State Water Resources Control Board (2016). Frequently Asked Questions About Recycled Oilfield Water for Crop Irrigation. https://www.waterboards.ca.gov/publications_forms/publications/factsheets/docs/prod_water_for_crop_irrigation.pdf

[13] North Dakota Department of Health – Guidelines for the Use of Oilfield Salt Brines for Dust and Ice Control. https://deq.nd.gov/Publications/WQ/1_GW/general/IceDustControlUsingOilfieldBrine_20130321.pdf

[14] 1509.226: Surface applications of brine by local governments. Ohio Revised Code, Title XV: Conservation of Natural Resources, Chapter 1509: Division of Oil and Gas Resources Management – Oil and Gas. http://codes.ohio.gov/orc/1509.226

[15] Poole, H. (2013). State Policies on Use of Hydraulic Fracturing Waste as a Road Deicer. Connecticut Office of Legislative Research. https://www.cga.ct.gov/2013/rpt/2013-R-0469.htm

[16] Schnebele, E. (2016). Iodine. 2015 Minerals Yearbook, U.S. Geological Survey. https://minerals.usgs.gov/minerals/pubs/commodity/iodine/myb1-2015-iodin.pdf

Part 6: Groundwater Protection in Oil and Gas Production

[1] Maupin, M.A. et al. (2014). Estimated Water Use in the United States in 2010, U.S. Geological Survey Circular 1405. https://pubs.usgs.gov/circ/1405/

[2] U.S. Geological Survey – The USGS Water Science School: Aquifers and Groundwater. https://water.usgs.gov/edu/earthgwaquifer.html

[3] Alley, W.M. et al. (1999). Sustainability of Ground-Water Resources. U.S. Geological Survey Circular 1186. https://pubs.usgs.gov/circ/circ1186/

[4] Ressetar, R. (2012). Energy News: Hydraulic Fracturing and Shale Gas. Utah Geological Survey, Survey Notes, 44(2), May 2012. https://geology.utah.gov/map-pub/survey-notes/energy-news/energy-news-hydraulic-fracturing-and-shale-gas/

Petroleum and the Environment
References

5. FracFocus Chemical Disclosure Registry – Chemical Use in Hydraulic Fracturing. http://fracfocus.org/water-protection/drilling-usage
6. FracFocus Chemical Disclosure Registry. http://fracfocus.org/
7. "DEP: Unauthorized drilling fluid contaminated Potter County Aquifer." S. Phillips, NPR StateImpact Pennsylvania, October 1, 2015. https://stateimpact.npr.org/pennsylvania/2015/10/01/dep-unauthorized-drilling-fluid-contaminated-potter-county-aquifer/
8. National Ground Water Association (2013). Water Wells in Proximity to Natural Gas or Oil Development. NGWA Information Brief, updated September 2017. http://www.ngwa.org/Media-Center/briefs/Documents/Info-Brief-Hydraulic-Fracturing.pdf
9. Kell, S. (2011). State Oil and Gas Agency Groundwater Investigations and Their Role in Advancing Regulatory Reforms, A Two State Review: Ohio and Texas. Ground Water Protection Council, August 2011. http://www.gwpc.org/sites/default/files/State%20Oil%20%26%20Gas%20Agency%20Groundwater%20Investigations.pdf
10. U.S. Geological Survey – U.S. Geological Survey Identifies Crude-Oil Metabolites in Subsurface Plumes. https://toxics.usgs.gov/highlights/2016-03-31-crude_oil_metabolites.html
11. FracFocus Chemical Disclosure Registry – Hydraulic Fracturing: The Process. https://fracfocus.org/hydraulic-fracturing-how-it-works/hydraulic-fracturing-process
12. U.S. Environmental Protection Agency (2016). Hydraulic Fracturing for Oil and Gas: Impacts from the Hydraulic Fracturing Water Cycle on Drinking Water Resources in the United States (Final Report). EPA/600/R-16/236F. https://cfpub.epa.gov/ncea/hfstudy/recordisplay.cfm?deid=332990
13. Wyoming Department of Environmental Quality – Pavillion Investigation. http://deq.wyoming.gov/wqd/pavillion-investigation/
14. "Last two Dimock families settle lawsuit with Cabot over water." J. Hurdle, NPR StateImpact Pennsylvania, September 26, 2017. https://stateimpact.npr.org/pennsylvania/2017/09/26/last-two-dimock-families-settle-lawsuit-with-cabot-over-water/
15. Railroad Commission of Texas (2014). Water Well Complaint Investigation Report, Silverado on the Brazos Neighborhood, Parker County, Texas, May 23, 2014. Accessed at: https://assets.documentcloud.org/documents/1175488/water-well-complaint-investigation-report-5-23.pdf
16. Patterson, L.A. et al. (2017). Unconventional Oil and Gas Spills: Risks, Mitigation Priorities, and State Reporting Requirements. Environ. Sci. Technol., 51(5), 2563-2573. https://pubs.acs.org/doi/abs/10.1021/acs.est.6b05749
17. Commonwealth of Pennsylvania – The Pennsylvania Code: § 78.84. Casing standards; and §78.85. Cement standards. In: Chapter 78, Subchapter D: Well Drilling, Operation, and Plugging. https://www.pacode.com/secure/data/025/chapter78/subchapDtoc.html
18. Sherwood, O.A. et al. (2016). Groundwater methane in relation to oil and gas development and shallow coal seams in the Denver-Julesburg Basin of Colorado. Proc. Natl. Acad. Sci. U.S.A., 113(30), 8391-8396. http://www.pnas.org/content/113/30/8391
19. Darrah, T.H. et al. (2014). Noble gases identify the mechanisms of fugitive gas contamination in drinking-water wells overlying the Marcellus and Barnett Shales. Proc. Natl. Acad. Sci. U.S.A., 111(39), 14076-14081. http://www.pnas.org/content/111/39/14076
20. Ohio Department of Natural Resources, Division of Mineral Resources Management (2008). Report on the Investigation of the Natural Gas Invasion of Aquifers in Bainbridge Township of Geauga County, Ohio. http://oilandgas.ohiodnr.gov/portals/oilgas/pdf/bainbridge/report.pdf
21. Boyer, E.W. et al. (2012). The Impact of Marcellus Gas Drilling on Rural Drinking Water Supplies, Final report to the Center for Rural Pennsylvania, March 2012. http://www.rural.palegislature.us/documents/reports/Marcellus_and_drinking_water_2012.pdf
22. Baldassare, F. et al. (2014). A geochemical context for stray gas investigations in the northern Appalachian Basin: Implications of analyses of natural gases from Neogene-through Devonian-age strata. AAPG Bull., 98(2), 341-372. https://pubs.geoscienceworld.org/aapgbull/article-abstract/98/2/341/133388/a-geochemical-context-for-stray-gas-investigations
23. McMahon, P.B. et al. (2017). Methane and Benzene in Drinking-Water Wells Overlying the Eagle Ford, Fayetteville, and Haynesville Shale Hydrocarbon Producing Areas. Environ. Sci. Technol., 51(12), 6727-6734. https://pubs.acs.org/doi/abs/10.1021/acs.est.7b00746

Part 7: Abandoned Wells

1. U.S. Energy Information Administration – U.S. Oil and Natural Gas Wells by Production Rate. https://www.eia.gov/petroleum/wells/
2. National Petroleum Council (2011). Plugging and Abandonment of Oil and Gas Wells. In: Prudent Development: Realizing the Potential of North America's Abundant Natural Gas and Oil Resources, Topic Paper #2-25. https://www.npc.org/Prudent_Development-Topic_Papers/2-25_Well_Plugging_and_Abandonment_Paper.pdf
3. Orphan Well Association – Frequently Asked Questions. http://www.orphanwell.ca/pg_faq.html
4. U.S. General Accounting Office (1989). Drinking Water: Safeguards Are Not Preventing Contamination from Injected Oil and Gas Wastes. GAO/RCED-89-97. https://www.gao.gov/assets/150/147952.pdf
5. Image source: https://pixnio.com/miscellaneous/abandoned-gas-well-pump
6. Brandt, A.R. et al. (2014). Methane Leaks from North American Natural Gas Systems. Science, 343, 733-735. http://science.sciencemag.org/content/343/6172/733.full
7. Ground Water Protection Council (2011). State Oil and Gas Agency Groundwater Investigations and Their Role in Advancing Regulatory Reforms – A Two-State Review: Ohio and Texas. http://www.gwpc.org/sites/default/files/State%20Oil%20%26%20Gas%20Agency%20Groundwater%20Investigations.pdf
8. Townsend-Small, A. et al. (2016). Emissions of coalbed and natural gas methane from abandoned oil and gas wells in the United States. Geophys. Res. Lett., 43, 2283-2290. https://agupubs.onlinelibrary.wiley.com/doi/full/10.1002/2015GL067623
9. "Well near Berthoud starts spilling drilling mud 33 years after it was capped." P. Johnson, Denver Post, October 31, 2017. https://www.denverpost.com/2017/10/31/well-near-berthoud-starts-spilling-drilling-mud-33-years-after-capped/
10. U.S. Environmental Protection Agency (2016). Hydraulic Fracturing for Oil and Gas: Impacts from the Hydraulic Fracturing Water Cycle on Drinking Water Resources in the United States. https://cfpub.epa.gov/ncea/hfstudy/recordisplay.cfm?deid=332990

Petroleum and the Environment References

[11] Interstate Oil & Gas Compact Commission (2016). State Financial Assurance Requirements. http://iogcc.ok.gov/Websites/iogcc/images/Financial_Assurances_FINAL_web.pdf

[12] Interstate Oil & Gas Compact Commission – State orphaned and abandoned well programs. http://groundwork.iogcc.ok.gov/node/428

[13] Oklahoma Energy Resources Board – Well Site Clean Up. https://www.oerb.com/well-site-clean-up

[14] Pennsylvania Department of Environmental Protection (2017). Abandoned and Orphan Oil and Gas Wells and the Well Plugging Program. Fact Sheet 8000-FS-DEP1670. http://www.depgreenport.state.pa.us/elibrary/GetDocument?docId=3823&DocName=8000-FS-DEP1670.pdf

[15] Texas Comptroller – Crude Oil Production Tax. https://comptroller.texas.gov/taxes/crude-oil/

[16] Oklahoma Energy Resources Board – Funding. https://oerb.com/about/funding

[17] Interstate Oil & Gas Compact Commission (2009), Orphaned and Abandoned Wells: Innovative Solutions. Groundwork, October 2009. http://groundwork.iogcc.ok.gov/sites/default/files/Orphaned%20Wells%20Case%20Study_0.pdf

[18] Texas Railroad Commission (2017). Monthly Report of State-funded Well Plugging Activities, August 2017, FY 2017. http://www.rrc.state.tx.us/media/41910/aug17-sfp-report.pdf

[19] Texas Railroad Commission (2017). Monthly Report of State-funded Well Plugging Activities, December 2017, FY 2018. http://www.rrc.state.tx.us/media/43398/dec-17-sfp-report.pdf

[20] Pennsylvania Department of Environmental Protection – 2016 Oil and Gas Annual Report. http://www.depgis.state.pa.us/oilgasannualreport/index.html#legacy

[21] Oklahoma Energy Resources Board – Restoring Oklahoma's Land. https://oerb.com/uploads/17oer10848-es-fact-sheet-land-restoring-ok-land.pdf

[22] "Orphan Oil Wells in L.A. Neighborhood to be Permanently Sealed." California Department of Conservation press release, June 8, 2016. http://www.conservation.ca.gov/index/news/Documents/2016-13%20Los%20Angeles%20orphan%20well%20plugging.pdf

[23] U.S. Government Accountability Office (2011). Oil and Gas Bonds: BLM Needs a Comprehensive Strategy to Better Manage Potential Oil and Gas Well Liability. GAO-11-292. https://www.gao.gov/new.items/d11292.pdf

Part 8: What Determines the Location of a Well?

[1] Biewick, L.R.H. (2008). Areas of Historical Oil and Gas Exploration and Production in the United States. U.S. Geological Survey Digital Data Series DDS-69-Q. https://pubs.usgs.gov/dds/dds-069/dds-069-q/text/pdfmaps.htm

[2] U.S. Energy Information Administration – Lower 48 states shale plays, June 2016. https://www.eia.gov/maps/images/shale_gas_lower48.jpg

[3] U.S. Energy Information Administration (2017). Marcellus Shale Play: Geology Review. https://www.eia.gov/maps/pdf/MarcellusPlayUpdate_Jan2017.pdf

[4] U.S. Energy Information Administration – Marcellus and Utica/Point Pleasant wells through April 2017. https://www.eia.gov/maps/images/Marcellus_UticaPointPleasant_Wells_April2017.jpg

[5] U.S. Bureau of Land Management – Oil and Gas: Leasing. https://www.blm.gov/programs/energy-and-minerals/oil-and-gas/leasing

[6] Pennsylvania Department of Conservation and Natural Resources (2017). Natural Gas Development and State Forests: Shale Gas Leasing Statistical Summary, May 2017. http://www.docs.dcnr.pa.gov/cs/groups/public/documents/document/dcnr_20029363.pdf

[7] U.S. Bureau of Indian Affairs – Working on Indian Lands. https://www.bia.gov/as-ia/ieed/division-energy-and-mineral-development/working-indian-lands

[8] Joy, M.P. and Dimitroff, S.D. (2016). Oil and gas regulation in the United States: overview. Westlaw, June 1, 2016. https://content.next.westlaw.com/Document/I466099551c9011e38578f7ccc38dcbee/View/FullText.html

[9] Bosquez IV, T. et al. (2015). Fracking Debate: The Importance of Pre-Drill Water-Quality Testing. American Bar Association, Section of Litigation: Environmental Litigation. http://apps.americanbar.org/litigation/committees/environmental/articles/winter2015-0215-fracking-debate-importance-pre-drill-water-quality-testing.html

[10] New York Department of Environmental Conservation – High-Volume Hydraulic Fracturing in New York State. https://www.dec.ny.gov/energy/75370.html

[11] General Assembly of Maryland – HB1325 (CH0013): Oil and Natural Gas – Hydraulic Fracturing – Prohibition. Approved by the Governor, April 4, 2017. http://mgaleg.maryland.gov/webmga/frmMain.aspx?pid=billpage&stab=01&id=hb1325&tab=subject3&ys=2017RS

[12] Vermont Department of Environmental Conservation (2015). A Report on the Regulation and Safety of Hydraulic Fracturing for Oil or Natural Gas Recovery. https://legislature.vermont.gov/assets/Legislative-Reports/ANR-REPORT-REGULATION-OF-HF-FOR-OIL-OR-NATURAL-GAS-RECOVERY-2015.02.12.FINAL.pdf

[13] "BLM Lands Leasing." Statement of Neil Kornze, Director, Bureau of Land Management, U.S. Department of the Interior, before the House Committee on Oversight and Government Reform, March 23, 2016. https://www.doi.gov/ocl/blm-lands-leasing

[14] U.S. Bureau of Land Management – Oil and Gas Statistics: Table 8 – Wells Spud. https://www.blm.gov/programs/energy-and-minerals/oil-and-gas/oil-and-gas-statistics

[15] U.S. Army Corps of Engineers (2017). U.S. Army Corps of Engineers Regulatory Role in Activities Associated with Oil and Natural Gas Production and Distribution, June 2017. http://www.lrb.usace.army.mil/Portals/45/docs/regulatory/DistrictInfo/NY_Oil_and_Gas_FactSheet_06-2017.pdf?ver=2017-07-24-104055-073

[16] U.S. Environmental Protection Agency – What is the National Environmental Policy Act? https://www.epa.gov/nepa/what-national-environmental-policy-act

[17] U.S. Bureau of Land Management – How We Manage. https://www.blm.gov/about/how-we-manage

[18] The Nature Conservancy – LEEP: The Nature Conservancy's Appalachian Shale Siting Tool. https://www.nature.org/ourinitiatives/regions/northamerica/areas/centralappalachians/leep-summary.pdf

[19] The Nature Conservancy and Carnegie Mellon University (2016). Advancing the Next Generation of Environmental Practices for Shale Development: Workshop Deliberations and Recommendations. May 27-29, 2015. Pittsburgh, PA. https://www.cmu.edu/energy/documents/Shale_Workshop_Deliberations_and_Recommendations_Final.pdf

[20] "Halliburton, Eclipse Resources complete longest lateral well in U.S." World Oil Magazine, May 31, 2016. http://www.worldoil.com/news/2016/5/31/halliburton-eclipse-resources-complete-longest-lateral-well-in-us

[21] U.S. Energy Information Administration – Crude Oil and Natural Gas Drilling Activity. https://www.eia.gov/dnav/ng/ng_enr_drill_s1_m.htm

Petroleum and the Environment References

[22] U.S. Department of Energy – Enhanced Oil Recovery. https://www.energy.gov/fe/science-innovation/oil-gas-research/enhanced-oil-recovery

[23] "What is "Forced" Pooling and Why is it Important?" J. Luellen, Husch Blackwell Emerging Energy Insights, April 25, 2017. https://www.emergingenergyinsights.com/2017/04/forced-pooling-important/

[24] U.S. Energy Information Administration (2016). EIA report shows decline in cost of U.S. oil and gas wells since 2012. Today in Energy, March 30, 2016. https://www.eia.gov/todayinenergy/detail.php?id=25592

[25] Railroad Commission of Texas (2017). Summary of Drilling, Completion and Plugging Reports Processed for 2016. http://www.rrc.state.tx.us/media/37649/annual2016.pdf

Part 9: Land Use in the Oil and Gas Industry

[1] State of California – § 1776. Well Site and Lease Restoration, California Statutes and Regulations for the Division of Oil, Gas, & Geothermal Resources, January 2018. ftp://ftp.consrv.ca.gov/pub/oil/laws/PRC10.pdf

[2] Kansas Office of Revisor of Statutes – Kansas Statutes Chapter 55: Oil and Gas, Article 1: Oil and Gas Wells; Regulatory Provisions, 55-177: Removal of structures and abutments from lands after abandoning wells; exception. https://www.ksrevisor.org/statutes/ksa_ch55.html

[3] Commonwealth of Pennsylvania – The Pennsylvania Code: § 78.a.65. Site restoration. https://www.pacode.com/secure/data/025/chapter78a/s78a.65.html

[4] Bureau of Land Management (2007). Chapter 6 – Reclamation and Abandonment. In: Surface Operating Standards and Guidelines for Oil and Gas, Fourth Edition – Revised 2007. https://www.blm.gov/sites/blm.gov/files/Chapter%206%20-%20Reclamation%20and%20Abandonment.pdf

[5] Penn State Public Broadcasting – Explore Shale: How much land in Pennsylvania has been affected by Marcellus Shale drilling? Based on data from FracTracker Alliance and the Pennsylvania Department of Environmental Protection. http://exploreshale.org/

[6] U.S. Government Accountability Office (2014). Updated Guidance, Increased Coordination, and Comprehensive Data Could Improve BLM's Management and Oversight. https://www.gao.gov/products/GAO-14-238

[7] "Halliburton, Eclipse Resources complete longest lateral well in U.S." World Oil Magazine, May 31, 2016. http://www.worldoil.com/news/2016/5/31/halliburton-eclipse-resources-complete-longest-lateral-well-in-us

[8] Cochener, J. (2010). Quantifying Drilling Efficiency. Working Paper, U.S. Energy Information Administration. https://www.eia.gov/workingpapers/pdf/drilling_efficiency.pdf

[9] Railroad Commission of Texas – Oil and Gas Well Records. http://www.rrc.state.tx.us/oil-gas/research-and-statistics/obtaining-commission-records/oil-and-gas-well-records/

[10] U.S. Energy Information Administration – Natural Gas Annual Respondent Query System (EIA-757 Data through 2014): 757 Processing Capacity. https://www.eia.gov/cfapps/ngqs/ngqs.cfm?f_report=RP9&f_sortby=&f_items=&f_year_start=&f_year_end=&f_show_compid=&f_fullscreen

[11] U.S. Energy Information Administration – Number and Capacity of Petroleum Refineries. https://www.eia.gov/dnav/pet/pet_pnp_cap1_a_(na)_8O0_Count_a.htm

[12] McDonald, R.I. et al. (2009). Energy Sprawl or Energy Efficiency: Climate Policy Impacts on Natural Habitat for the United States of America. PLoS ONE 4(8): e6802. http://journals.plos.org/plosone/article?id=10.1371/journal.pone.0006802

Part 10: The Pinedale Gas Field, Wyoming

[1] Wyoming State Geological Survey – Cultural Geology Guide: Pinedale Anticline. http://www.wsgs.wyo.gov/public-info/guide-pinedale

[2] U.S. Bureau of Land Management (2008). Record of Decision for the Supplemental Environmental Impact Statement for the Pinedale Anticline Oil and Gas Exploration and Development Project, Sublette County, Wyoming. https://eplanning.blm.gov/epl-front-office/eplanning/planAndProjectSite.do?methodName=dispatchToPatternPage&currentPageId=88620

[3] "Wyoming natural gas production falls for sixth consecutive year; U.S. production rises." B. Storrow, Casper Star Tribune, January 5, 2016. http://trib.com/business/energy/wyoming-natural-gas-production-falls-for-sixth-consecutive-year-u/article_d715399e-09ca-52a0-8db9-594171d3edd1.html

[4] U.S. Energy Information Administration (2017). Marcellus Region: Drilling Productivity Report, July 2017. https://www.eia.gov/petroleum/drilling/pdf/marcellus.pdf

[5] AAPG Wiki – Pinedale field. http://wiki.aapg.org/Pinedale_field. Image reproduced according to a CC BY-SA 3.0 license: https://creativecommons.org/licenses/by-sa/3.0/

[6] U.S. Fish and Wildlife Service – Greater Sage-Grouse. https://www.fws.gov/greatersagegrouse/

[7] Hayden-Wing Associates, LLC (2016). Pygmy Rabbit Monitoring in the Pinedale Anticline Project Area, Sublette County, Wyoming, 2016. Report prepared for Wyoming Game and Fish Department, Pinedale Anticline Project Office, and Bureau of Land Management, Pinedale, Wyoming. https://www.blm.gov/sites/blm.gov/files/documents/files/PAPA_PygmyRabbit_Report_2016_WordDocument01122017_0.pdf

[8] Sawyer, H. et al. (2017). Mule deer and energy development – Long-term trends of habituation and abundance. Glob. Chang. Biol., 23, 4521-4529. https://onlinelibrary.wiley.com/doi/abs/10.1111/gcb.13711

[9] U.S. Bureau of Land Management (1987). Pinedale Resource Area: proposed resource management plan / environmental impact statement: final. https://archive.org/details/pinedaleresourceunit

[10] U.S. Bureau of Land Management (1989). Record of decision and resource management plan for the Pinedale Resource Area. https://archive.org/details/recordofdecision27unit

[11] U.S. Bureau of Land Management (2008). Pinedale Resource Management Plan. https://eplanning.blm.gov/epl-front-office/eplanning/planAndProjectSite.do?methodName=dispatchToPatternPage&currentPageId=88620

[12] Except where otherwise referenced, information in this section comes from communications with BLM Pinedale Anticline Project Office staff, June 30, 2017.

Petroleum and the Environment References

13. Boschee, P. (2012). Handling Produced Water from Hydraulic Fracturing. Oil and Gas Facilities Magazine, 1 (1). https://www.spe.org/en/ogf/ogf-article-detail/?art=364
14. Shafer, L. (2011). Water Recycling and Purification in the Pinedale Anticline Field: Results from the Anticline Disposal Project. SPE Americas E&P Health, Safety, Security, and Environmental Conference, 21-13 March, Houston, Texas, USA. https://www.onepetro.org/conference-paper/SPE-141448-MS
15. Wyoming Department of Environmental Quality (2015). Proposed Revisions to the Chapter 6, Section 2 Oil and Gas Production Facilities Permitting Guidance. Technical Support Document. http://deq.wyoming.gov/media/attachments/Air%20Quality/Rule%20Development/Proposed%20Rules%20and%20Regulations/Oil-and-Gas-Guidance-Revision_Technical-Support-Document_Draft-9-24-2015.pdf
16. U.S. Environmental Protection Agency (2016). Determinations of Attainment by the Attainment Date, Extensions of the Attainment Date, and Reclassification of Several Areas for the 2008 Ozone National Ambient Air Quality Standards. Federal Register, 81 (86), p. 26697-26722, May 4, 2016. https://www.gpo.gov/fdsys/pkg/FR-2016-05-04/pdf/2016-09729.pdf

Part 11: Heavy Oil

1. U.S. Geological Survey (2003). Heavy Oil and Natural Bitumen: Strategic Petroleum Resources. https://pubs.usgs.gov/fs/fs070-03/fs070-03.pdf
2. Natural Resources Canada – Environmental Challenges. https://www.nrcan.gc.ca/energy/oil-sands/5855
3. File:Aurora – tar sands.png. Wikimedia Commons user Int23. https://commons.wikimedia.org/wiki/File:Aurora_-_tar_sands.png. Reproduced according to a CC BY-SA 4.0 license: https://creativecommons.org/licenses/by-sa/4.0/deed.en
4. Natural Resources Canada – Oil Sands Extraction and Processing. http://www.nrcan.gc.ca/energy/oil-sands/18094
5. American Association of Petroleum Geologists, Energy Minerals Division (2015). Unconventional Energy Resources: 2015 Review. Natural Resources Research, 24, 443-508. https://link.springer.com/article/10.1007/s11053-015-9288-6
6. Koottungal, L. (2014). 2014 worldwide EOR survey. Oil & Gas Journal, 112(4), 78-91. https://www.ogj.com/articles/print/volume-112/issue-4/special-report-eor-heavy-oil-survey/2014-worldwide-eor-survey.html
7. Society of Petroleum Engineers PetroWiki – Cold heavy oil production with sand. http://petrowiki.org/Cold_heavy_oil_production_with_sand
8. Carnegie Endowment for International Peace – Oil-Climate Index: Profiling Emissions in the Supply Chain. http://oci.carnegieendowment.org/#supply-chain
9. Gosselin, P. et al. (2010). Environmental and Health Impacts of Canada's Oil Sands Industry. The Royal Society of Canada Expert Health Panel, December 2010. https://rsc-src.ca/en/expert-panels/rsc-reports/environmental-and-health-impacts-canadas-oil-sands-industry
10. Frank, R.A. et al. (2014). Profiling Oil Sands Mixtures from Industrial Developments and Natural Groundwaters for Source Identification. Env. Sci. Technol., 48(%), 2660-2670. https://pubs.acs.org/doi/abs/10.1021/es500131k
11. Roy, J.W. et al. (2016). Assessing Risks of Shallow Riparian Groundwater Quality Near an Oil Sands Tailings Pond. Groundwater, 54(4), 545-558. https://onlinelibrary.wiley.com/doi/abs/10.1111/gwat.12392
12. Government of Canada – Monitoring air quality in Alberta oil sands. https://www.canada.ca/en/environment-climate-change/services/oil-sands-monitoring/monitoring-air-quality-alberta-oil-sands.html
13. Government of Canada (2015). Oil Sands: Land Use and Reclamation. Factsheet, May 2015. http://www.nrcan.gc.ca/sites/www.nrcan.gc.ca/files/energy/pdf/eneene/pubpub/pdf/os2015/14-0702-Oil-Sands-Land-Use-and-Reclamation_access_eng.pdf
14. Tar Sands, Alberta. Howl Arts Collective, Flickr. https://www.flickr.com/photos/67952496@N05/6544064931. Reproduced according to a CC BY 2.0 license https://creativecommons.org/licenses/by/2.0/
15. Canada National Energy Board – Commodity Statistics, Summary Crude Oil Export by Type. https://apps.neb-one.gc.ca/CommodityStatistics/Statistics.aspx?language=english
16. U.S. Energy Information Administration – Petroleum & Other Liquids: U.S. Imports by Country of Origin. https://www.eia.gov/dnav/pet/pet_move_impcus_a2_nus_epc0_im0_mbblpd_a.htm

Part 12: Oil and Gas in the U.S. Arctic

1. U.S. Geological Survey (2008). Circum-Arctic Resource Appraisal: Estimated of Undiscovered Oil and Gas North of the Arctic Circle. Fact Sheet 2008-3049. https://pubs.usgs.gov/fs/2008/3049/fs2008-3049.pdf
2. Alaska Department of Transportation & Public Facilities – Arctic Port Study. http://www.dot.state.ak.us/stwddes/desports/arctic.shtml
3. Comay, L.B. et al. (2018). Arctic National Wildlife Refuge (ANWR): An Overview. Congressional Research Service, 7-5700, January 9, 2018.
4. U.S. Bureau of Land Management – National Petroleum Reserve in Alaska. https://www.blm.gov/programs/energy-and-minerals/oil-and-gas/about/alaska/NPR-A
5. Attanasi, E.D. (2003). Economics of Undiscovered Oil in the Federal Lands of the National Petroleum Reserve, Alaska. U.S. Geological Survey Open-File Report 03-044. https://pubs.usgs.gov/of/2003/of03-044/
6. U.S. Geological Survey (2017). Assessment of Undiscovered Oil and Gas Resources in the Cretaceous Nanushuk and Torok Formations, Alaska North Slope, and Summary of Resource Potential of the National Petroleum Reserve in Alaska, 2017. Fact Sheet 2017-3088. https://pubs.usgs.gov/fs/2017/3088/fs20173088.pdf
7. U.S. Fish and Wildlife Service – Arctic: National Wildlife Refuge-Alaska. https://www.fws.gov/refuge/arctic/facts_and_features.html
8. U.S. Geological Survey – Arctic National Wildlife Refuge, 1002 Area, Petroleum Assessment, 1998 (revised), Including Economic Analysis. Fact Sheet 0028-01: Online Report. https://pubs.usgs.gov/fs/fs-0028-01/fs-0028-01.htm

Petroleum and the Environment References

9. H.R. 1 – An Act to provide for reconciliation pursuant to titles II and V of the concurrent resolution on the budget for fiscal year 2018. 115th Congress (2017-2018). https://www.congress.gov/bill/115th-congress/house-bill/1
10. U.S. Bureau of Ocean Energy Management – Outer Continental Shelf. https://www.boem.gov/Outer-Continental-Shelf/
11. Alaska Oil and Gas Conservation Commission – Northstar Oil Pool Statistics. http://doa.alaska.gov/ogc/annual/current/18_oil_pools/northstar-%20oil/1_oil_1.htm
12. Bureau of Ocean Energy Management – 2019-2024 National Outer Continental Shelf Oil and Gas Leasing Program. https://www.boem.gov/National-Program/
13. Alaska Oil and Gas Conservation Commission. http://doa.alaska.gov/ogc/
14. U.S. Bureau of Safety and Environmental Enforcement – Fact Sheet: Arctic Drilling Rule. https://www.bsee.gov/guidance-and-regulations/regulations/arctic-rule
15. The Pew Charitable Trusts (2013). Arctic Standards: Recommendations on Oil Spill Prevention, Response, and Safety in the U.S. Arctic Ocean. http://www.pewtrusts.org/~/media/assets/2013/09/23/arcticstandardsfinal.pdf
16. "The Only Safe Arctic Drilling is No Arctic Drilling." Natural Resources Defense Council Press Release, July 7, 2016. https://www.nrdc.org/media/2016/160707
17. National Research Council (2003). Cumulative Environmental Effects of Oil and Gas Activities on Alaska's North Slope. Washington, DC: The National Academies Press. https://www.nap.edu/catalog/10639/cumulative-environmental-effects-of-oil-and-gas-activities-on-alaskas-north-slope
18. Alaska Oil and Gas Conservation Commission – Colville River Unit, Alpine Oil Pool. http://doa.alaska.gov/ogc/annual/current/18_oil_pools/colville%20river%20-%20oil/colville%20river,%20alpine%20-%20oil/1_oil_1.htm
19. Alaska Oil and Gas Conservation Commission – Production. http://aogweb.state.ak.us/DataMiner3/Forms/Production.aspx
20. U.S. Bureau of Land Management (2015). Supplemental Environmental Impact Statement for the Alpine Satellite Development Plan for the Proposed Greater Mooses Tooth One Development Project: Record of Decision, February 2015. https://eplanning.blm.gov/epl-front-office/projects/nepa/37035/54639/59351/MASTER_GMT1ROD.Ver17_signed__2.13.15.pdf
21. Alyeska Pipeline Service Company – Safety & Environment: Permafrost. http://www.alyeska-pipe.com/SafetyEnvironment/EnvironmentalProtection/Permafrost
22. Alyeska Pipeline Service Company – Trans Alaska Pipeline System Facts. http://www.alyeska-pipe.com/assets/uploads/pagestructure/NewsCenter_MediaResources_FactSheets_Entries/635078372894251917_2013AlyeskaTAPSFactBook.pdf
23. Alyeska Pipeline Service Company – Earthquake Protection. http://www.alyeska-pipe.com/TAPS/HistoryDesignConstruction/EarthquakeProtection
24. U.S. Geological Survey – Denali Fault: Alaska Pipeline. https://www.usgs.gov/media/images/denali-fault-alaska-pipeline
25. Nuka Research and Planning Group (2013). Oil Spill Occurrence Rates for Alaska North Slope Crude & Refined Oil Spills. Report to the Bureau of Ocean Energy Management, October 2013. OCS Study BOEM 2013-205. https://www.boem.gov/uploadedFiles/BOEM/BOEM_Newsroom/Library/Publications/131104_BOEMOilSpillOccurrenceFinalReport.pdf
26. Alaska Department of Environmental Conservation – PRP Spills Database Search. http://dec.alaska.gov/Applications/SPAR/PublicMVC/PERP/SpillSearch
27. "BP Exploration Alaska to Pay $25 Million Penalty for Alaskan North Slope Oil Spill." U.S. Department of Justice, May 3, 2011. https://www.justice.gov/opa/pr/bp-exploration-alaska-pay-25-million-penalty-alaskan-north-slope-oil-spill
28. Alaska Department of Environmental Conservation – GC-2 Transit Line Spill: Tundra Treatment Plan Rev.1, March 18, 2006. http://dec.alaska.gov/spar/ppr/response/sum_fy06/060302301/plans/060302301_plan_tundratreat0318.pdf
29. U.S. Bureau of Safety and Environmental Enforcement – Arctic Oil Spill Response Research. https://www.bsee.gov/site-page/arctic-oil-spill-response-research
30. Alaska Department of Environmental Conservation – 20 Years after the Exxon Valdez. http://dec.alaska.gov/spar/evos/thennow.htm
31. U.S. National Oceanic and Atmospheric Administration – 25 Years Later: Timeline of Recovery from Exxon Valdez Oil Spill. https://response.restoration.noaa.gov/oil-and-chemical-spills/significant-incidents/exxon-valdez-oil-spill/timeline-ecological-recovery-infographic.html
32. U.S. Bureau of Safety and Environmental Enforcement – International Forums: The Arctic Offshore Regulators Forum (AORF). https://www.bsee.gov/what-we-do/international-engagement/forums
33. U.S. Energy Information Administration – Alaska North Slope Crude Oil Production. https://www.eia.gov/dnav/pet/hist/LeafHandler.ashx?n=pet&s=manfpak2&f=a

Part 13: Offshore Oil and Gas

1. U.S. Energy Information Administration (2016). Trends in U.S. Oil and Natural Gas Upstream Costs, March 2016. https://www.eia.gov/analysis/studies/drilling/pdf/upstream.pdf
2. U.S. Energy Information Administration – Crude Oil Production. https://www.eia.gov/dnav/pet/pet_crd_crpdn_adc_mbbl_m.htm
3. U.S. Energy Information Administration – Natural Gas Gross Withdrawals and Production https://www.eia.gov/dnav/ng/ng_prod_sum_a_EPG0_FGW_mmcf_a.htm
4. U.S. Department of the Interior – Natural Resources Revenue Data. https://revenuedata.doi.gov/
5. U.S. Energy Information Administration (2016). Offshore production nearly 30% of global crude oil output in 2015. Today in Energy, October 25, 2016. https://www.eia.gov/todayinenergy/detail.php?id=28492
6. U.S. Energy Information Administration (2016). Offshore oil production in deepwater and ultra-deepwater is increasing. Today in Energy, October 28, 2016. https://www.eia.gov/todayinenergy/detail.php?id=28552
7. Woods Hole Oceanographic Institution – Oil in the Ocean: FAQs. http://www.whoi.edu/oil/deepwater-horizon/faqs
8. U.S. Bureau of Ocean Energy Management (formerly Minerals Management Service) (2000). Deepwater Development: A Reference Document for the Deepwater Environmental Assessment, Gulf of Mexico OCS (1998 through 2007). MMS 2000-015. https://www.boem.gov/BOEM-Newsroom/Library/Publications/2000/2000-015.aspx
9. U.S. Energy Information Administration (2016). Crude oil prices started 2015 relatively low, ended the year lower. Today in Energy, January 6, 2016. https://www.eia.gov/todayinenergy/detail.php?id=24432

Petroleum and the Environment References

[10] U.S. Energy Information Administration – Monthly Crude Oil and Natural Gas Production, Crude Oil, Federal Offshore Gulf of Mexico. https://www.eia.gov/petroleum/production/

[11] "Department of Energy to Invest $30 Million to Boost Unconventional Oil and Natural Gas Recovery." U.S. Department of Energy, January 3, 2018. https://www.energy.gov/articles/department-energy-invest-30-million-boost-unconventional-oil-and-natural-gas-recovery

[12] DeepStar Program – About Us. https://deepstaraglobaloffshoreresearc.godaddysites.com/deepstar-program

[13] American Geosciences Institute (2016). Offshore Energy. https://www.americangeosciences.org/policy-critical-issues/webinars/offshore-energy

[14] "Seismic, Supercomputing Power Leads BP to More GoM Resources." V. Addison, Exploration and Production Magazine, April 28, 2017. https://www.epmag.com/seismic-supercomputing-power-leads-bp-more-gom-resources-1548411

[15] Dusseault, M.B. et al. (2004). Drilling Through Salt: Constitutive Behavior and Drilling Strategies, 6th North American Rock Mechanics Symposium, 5-9 June, Houston, Texas. https://www.onepetro.org/conference-paper/ARMA-04-608

[16] Shadravan, A. and Amani, M. (2012). HPHT 101: What Every Engineer or Geoscientist Should Know about High Pressure High Temperature Wells. SPE Kuwait International Petroleum Conference and Exhibition, 10-12 December, Kuwait City, Kuwait. https://www.onepetro.org/conference-paper/SPE-163376-MS

[17] "Drilling advances: Between a rock and a hot place." J. Redden, World Oil, vol. 237, no. 11, November 2016. http://www.worldoil.com/magazine/2016/november-2016/columns/drilling-advances

[18] Pai, S. (2016). Managing Risks around Rigs Using Autonomous Marine Vehicles. SPE/IADC Middle East Drilling Technology Conference and Exhibition, 26-28 January, Abu Dhabi, UAE. https://www.onepetro.org/conference-paper/SPE-178244-MS

[19] U.S. Global Change Research Program – 2014 National Climate Assessment: Changes in Hurricanes. https://nca2014.globalchange.gov/report/our-changing-climate/changes-hurricanes

[20] Transocean's Development Driller II. U.S. Coast Guard photo by Petty Officer 3rd Classy Barry Bena. Accessed via Wikimedia Commons. https://commons.wikimedia.org/wiki/File:Development-driller-2.jpg

[21] Clarke, K.C. and Hemphill, J.J. (2002). The Santa Barbara Oil Spill: A Retrospective. Yearbook of the Association of Pacific Coast Geographers, 64, 157-162. http://www.geog.ucsb.edu/~kclarke/Papers/SBOilSpill1969.pdf

[22] U.S. National Oceanic and Atmospheric Administration, Office of Response and Restoration – It Took More Than the Exxon Valdez Oil Spill to Pass the Historic Oil Pollution Act of 1990. http://response.restoration.noaa.gov/oil-and-chemical-spills/significant-incidents/exxon-valdez-oil-spill/it-took-more-exxon-valdez-oil-s

[23] National Academy of Engineering and National Research Council (2012). Macondo Well Deepwater Horizon Blowout: Lessons for Improving Offshore Drilling Safety. Washington, DC: The National Academies Press. https://www.nap.edu/catalog/13273/macondo-well-deepwater-horizon-blowout-lessons-for-improving-offshore-drilling

[24] U.S. Department of the Interior – Responsibilities Guide (BOEM & BSEE). https://www.boem.gov/A-to-Z-Guide-BOEM-BSEE-Functions/

[25] "U.S. and Five Gulf States Reach Historic Settlement with BP to Resolve Civil Lawsuit Over Deepwater Horizon Spill." U.S. Department of Justice, October 5, 2015. https://www.justice.gov/opa/pr/us-and-five-gulf-states-reach-historic-settlement-bp-resolve-civil-lawsuit-over-deepwater

[26] U.S. Bureau of Ocean Energy Management – BOEM Gulf of Mexico OCS Region Blocks and Active Leases by Planning Area, February 1, 2018. https://www.boem.gov/Gulf-of-Mexico-Region-Lease-Map/

[27] U.S. National Oceanic and Atmospheric Administration – Deepwater Horizon oil spill settlements: Where the money went. http://www.noaa.gov/explainers/deepwater-horizon-oil-spill-settlements-where-money-went

[28] Gulf of Mexico Research Institute: Investigating the effect of oil spills on the environment and public health. http://gulfresearchinitiative.org/

[29] "BSEE Responds to Oil Release in the Gulf of Mexico." U.S. Bureau of Safety and Environmental Enforcement, October 13, 2017. https://www.bsee.gov/newsroom/latest-news/statements-and-releases/press-releases/bsee-responds-to-oil-release-in-the-gulf

[30] "UPDATE 2: Coast Guard, federal agencies responding to offshore oil spill." U.S. Coast Guard, October 19, 2017. https://content.govdelivery.com/accounts/USDHSCG/bulletins/1beacc5

[31] U.S. Coast Guard (2015). Taylor Energy Oil Discharge at MC-20A Well Site and Ongoing Response Efforts. Incident Archive – Taylor Energy (Mississippi Canyon) Oil Spill – U.S. Coast Guard Fact Sheet, May 13, 2015. https://www.bsee.gov/newsroom/library/incident-archive/taylor-energy-mississippi-canyon/fact-sheet

[32] U.S. Bureau of Safety and Environmental Enforcement (2016). Oil and Gas and Sulfur Operations in the Outer Continental Shelf-Blowout Preventer Systems and Well Control. Federal Register, 81, p. 25887-26038, April 4, 2016. https://www.federalregister.gov/documents/2016/04/29/2016-08921/oil-and-gas-and-sulfur-operations-in-the-outer-continental-shelf-blowout-preventer-systems-and-well

[33] U.S. Bureau of Ocean Energy Management – Development of a New National OCS Program (2019-2024). https://www.boem.gov/National-OCS-Program/

Part 14: Spills in Oil and Natural Gas Fields

[1] U.S. Pipeline and Hazardous Materials Safety Administration – Incident Statistics. https://www.phmsa.dot.gov/hazmat-program-management-data-and-statistics/data-operations/incident-statistics

[2] Patterson, L.A. et al. (2017). Unconventional Oil and Gas Spills: Risks, Mitigation Priorities, and State Reporting Requirements. Environ. Sci. Technol., 51(5), 2563-2573. https://pubs.acs.org/doi/abs/10.1021/acs.est.6b05749

[3] U.S. Federal Highway Administration – Freight Facts and Figures 2013: Table 5-18. Number and Volume of Oil Spills In and Around U.S. Waterways: 1990, 2000, and 2009-2011. https://ops.fhwa.dot.gov/freight/freight_analysis/nat_freight_stats/docs/13factsfigures/table5_18.htm

Petroleum and the Environment References

4. U.S. Bureau of Ocean Energy Management (2013). Oil Spill Occurrence Rates for Alaska North Slope Crude & Refined Oil Spills. Nuka Research & Planning Group, LLC, for BOEM, October 2013. OCS Study BOEM 2013-205. https://www.boem.gov/uploadedFiles/BOEM/BOEM_Newsroom/Library/Publications/131104_BOEMOilSpillOccurrenceFinalReport.pdf
5. National Oceanic and Atmospheric Administration – How Do Oil Spills out at Sea Typically Get Cleaned Up? https://response.restoration.noaa.gov/about/media/how-do-oil-spills-out-sea-typically-get-cleaned.html
6. Image from U.S. Bureau of Land Management. Accessed at: "Company faces $10K for spill near Green River." B. Maffly, The Salt Lake Tribune, September 4, 2015. http://archive.sltrib.com/article.php?id=2909655&itype=CMSID#gallery-carousel-446996
7. U.S. Environmental Protection Agency (2015). Review of State and Industry Spill Data: Characterization of Hydraulic Fracturing-Related Spills. https://www.epa.gov/sites/production/files/2015-05/documents/hf_spills_report_final_5-12-15_508_km_sb.pdf
8. Maloney, K.O. et al. (2016). Unconventional oil and gas spills: Materials, volumes and risks to surface waters in four states of the U.S.. Sci. Total Environ., 581-582, 369-377. https://www.sciencedirect.com/science/article/pii/S0048969716328327?via%3Dihub
9. FracFocus Chemical Disclosure Registry. http://fracfocus.org/
10. API Energy – Oil Spill Prevention and Response: Land. http://www.oilspillprevention.org/oil-spill-cleanup/land-oil-spill-cleanup
11. U.S. Fish & Wildlife Service (2010). Effects of Oil on Wildlife and Habitat. Fact Sheet, June 2010. https://www.fws.gov/home/dhoilspill/pdfs/DHJICFWSOilImpactsWildlifeFactSheet.pdf
12. U.S. Geological Survey – Marcellus Shale Storage Tanks. https://www.usgs.gov/media/images/marcellus-shale-storage-tanks
13. U.S. Environmental Protection Agency – Oil Spills Prevention and Preparedness Regulations. https://www.epa.gov/oil-spills-prevention-and-preparedness-regulations
14. U.S. Environmental Protection Agency (2015). Review of State and Industry Spill Data: Characterization of Hydraulic Fracturing-Related Spills. https://www.epa.gov/sites/production/files/2015-05/documents/hf_spills_report_final_5-12-15_508_km_sb.pdf
15. U.S. Federal Emergency Management Agency – Oil spill clean up in Kansas. https://www.fema.gov/media-library/assets/images/51499
16. API Energy – Oil Spill Prevention and Response: Land. http://www.oilspillprevention.org/oil-spill-cleanup/land-oil-spill-cleanup
17. API Energy – Oil Spill Prevention and Response: Wildlife Cleanup http://www.oilspillprevention.org/oil-spill-cleanup/oil-spill-cleanup-toolkit/wildlife-cleanup
18. Minnesota Pollution Control Agency – National Crude Oil Spill Research Site in Bemidji, Minnesota. https://www.pca.state.mn.us/waste/national-crude-oil-spill-research-site-bemidji-minnesota
19. API Energy – Oil Spill Prevention and Response: Toolkit. http://www.oilspillprevention.org/oil-spill-cleanup/oil-spill-cleanup-toolkit
20. U.S. Geological Survey – Produced Waters Database. https://eerscmap.usgs.gov/pwapp/
21. U.S. Geological Survey (2003). Environmental Impacts of Petroleum Production: Initial Results from the Osage-Skiatook Petroleum Environmental Research Sites, Osage County, Oklahoma. Water-Resources Investigation Report 03-4260. https://pubs.usgs.gov/wri/wri03-4260/
22. Cozzarelli, I.M. et al. (2017). Environmental signatures and effects of an oil and gas wastewater spill in the Williston Basin, North Dakota. Sci. Total Environ., 579, 1781-1793. https://www.sciencedirect.com/science/article/pii/S0048969716326201?via%3Dihub
23. U.S. Environmental Protection Agency – Blacktail Creek Spill, Marmon, ND, Region VIII. https://response.epa.gov/site/site_profile.aspx?site_id=9716
24. U.S. Environmental Protection Agency – Blacktail Creek Spill. https://response.epa.gov/site/image_zoom.aspx?site_id=9716&counter=234864

Part 15: Transportation of Oil, Gas, and Refined Products

1. U.S. Pipeline and Hazardous Materials Safety Administration – National Pipeline Mapping System. https://www.npms.phmsa.dot.gov/
2. U.S. Pipeline and Hazardous Materials Administration – Annual Report Mileage Summary Statistics. https://cms.phmsa.dot.gov/data-and-statistics/pipeline/annual-report-mileage-summary-statistics
3. American Petroleum Institute and Association of Oil Pipe Lines (2016). Pipeline Performance Report & Strategic Plan http://www.aopl.org/wp-content/uploads/2016/08/2016-API-AOPL-Annual-Pipeline-Safety-Excellence-Performance-Report-Strategic-Plan.pdf
4. U.S. Energy Information Administration – Petroleum and Other Liquids: Product Supplied. https://www.eia.gov/dnav/pet/pet_cons_psup_dc_nus_mbbl_a.htm
5. U.S. Energy Information Administration – Movements of Crude Oil and Selected Products by Rail. https://www.eia.gov/dnav/pet/pet_move_railNA_a_EPC0_RAIL_mbbl_a.htm
6. U.S. Energy Information Administration – Crude-by-rail volumes to the East Coast are declining. Today in Energy, August 3, 2016. https://www.eia.gov/todayinenergy/detail.php?id=27352
7. Canada National Energy Board – Canadian Crude Oil Exports by Rail – Monthly Data. https://www.neb-one.gc.ca/nrg/sttstc/crdlndptrlmprdct/stt/cndncrdlxprtsrl-eng.html
8. National Tank Truck Carriers (2015). Tank Truck Industry Market Analysis. http://www.tanktruck.org/publications/tank-truck-industry-market-analysis-now-available
9. The Brattle Group (2014). Understanding Crude Oil and Product Markets. Prepared for the American Petroleum Institute. http://www.api.org/~/media/files/oil-and-natural-gas/crude-oil-product-markets/crude-oil-primer/understanding-crude-oil-and-product-markets-primer-low.pdf
10. Texas A&M Transportation Institute, Center for Ports and Waterways (2017). A Modal Comparison of Domestic Freight Transportation Effects on the General Public, 2001 – 2014. Prepared for the National Waterways Foundation. http://nationalwaterwaysfoundation.org/documents/Final%20TTI%20Report%202001-2014%20Approved.pdf
11. File:Oil Barge at the Cape Cod Canal.jpg. Wikimedia Commons user Pvalerio. https://commons.wikimedia.org/wiki/File:Oil_Barge_at_the_Cape_Cod_canal.jpg. Reproduced according to a CC BY 3.0 license: https://creativecommons.org/licenses/by/3.0/deed.en
12. U.S. Energy Information Administration – U.S. Domestic Crude Oil Refinery Receipts by Barge. https://www.eia.gov/dnav/pet/hist/LeafHandler.ashx?n=PET&s=8_NA_8RBD_NUS_MBBL&f=A
13. U.S. Energy Information Administration – Monthly Energy Review: Petroleum, March 2018. https://www.eia.gov/totalenergy/data/monthly/index.php#petroleum
14. U.S. Energy Information Administration – Exports by Destination: Total Crude Oil and Products. https://www.eia.gov/dnav/pet/pet_move_expc_a_EP00_EEX_mbbl_a.htm

Petroleum and the Environment References

15. Natural Resources Canada – Frequently Asked Questions (FAQs) Concerning Federally-Regulated Petroleum Pipelines in Canada. http://www.nrcan.gc.ca/energy/infrastructure/5893#h-1-4
16. Canada National Energy Board – 2016 Oil Exports Statistics Summary. https://www.neb-one.gc.ca/nrg/sttstc/crdlndptrlmprdct/stt/crdlsmmr/2016/smmry2016-eng.html
17. U.S. Energy Information Administration (2018). 2018 Annual Energy Outlook: Natural Gas Imports and Exports. https://www.eia.gov/outlooks/aeo/data/browser/#/?id=76-AEO2018&region=0-0&cases=ref2018&start=2016&end=2050&f=A&linechart=ref2018-d121317a.3-76-AEO2018&sid=&sourcekey=0
18. U.S. Energy Information Administration – Petroleum and Other Liquids: U.S. Imports by Country of Origin, https://www.eia.gov/dnav/pet/pet_move_impcus_a2_nus_ep00_im0_mbblpd_a.htm
19. U.S. Energy Information Administration – Petroleum and Other Liquids: Exports by Destination. https://www.eia.gov/dnav/pet/pet_move_expc_a_EP00_EEX_mbblpd_a.htm
20. United Nations Conference on Trade and Development (2017). Review of Maritime Transport 2017. http://unctad.org/en/PublicationsLibrary/rmt2017_en.pdf
21. U.S. Energy Information Administration (2018). Natural gas prices, production, and exports increased from 2016-2017. Today in Energy, January 16, 2018. https://www.eia.gov/todayinenergy/detail.php?id=34532
22. U.S. Federal Energy Regulatory Commission – North American LNG Import/Export Terminals Approved (As of January 24, 2018). https://www.ferc.gov/industries/gas/indus-act/lng/lng-approved.pdf
23. U.S. Department of Energy (2018). LNG Monthly, January 2018. https://www.energy.gov/sites/prod/files/2018/03/f49/LNG%20Monthly%202018_2.pdf
24. U.S. Energy Information Administration – U.S. Natural Gas Exports and Re-Exports by Point of Exit. https://www.eia.gov/dnav/ng/ng_move_poe2_dcu_YENA-NJA_m.htm
25. U.S. Energy Information Administration – The Basics of Underground Natural Gas Storage. https://www.eia.gov/naturalgas/storage/basics/
26. U.S. Department of Energy – Strategic Petroleum Reserve. https://energy.gov/fe/services/petroleum-reserves/strategic-petroleum-reserve
27. U.S. Department of Energy – Northeast Home Heating Oil Reserve. https://energy.gov/fe/services/petroleum-reserves/heating-oil-reserve
28. U.S. Department of Energy – Northeast Gasoline Supply Reserve. https://energy.gov/fe/services/petroleum-reserves/northeast-regional-refined-petroleum-product-reserve
29. U.S. Energy Information Administration – Energy Explained: Use of Heating Oil https://www.eia.gov/energyexplained/index.cfm?page=heating_oil_use
30. U.S. Pipeline and Hazardous Materials Safety Administration – Office of Hazardous Materials Safety: Incident Reports Database Search. https://hazmatonline.phmsa.dot.gov/IncidentReportsSearch/
31. U.S. Pipeline and Hazardous Materials Safety Administration – Pipeline Incident 20 Year Trends: All Reported Incident 20 Year Trend. https://www.phmsa.dot.gov/data-and-statistics/pipeline/pipeline-incident-20-year-trends
32. U.S. Environmental Protection Agency – Extracting the Damaged Pipeline. Accessed on Flickr: https://www.flickr.com/photos/usepagov/4885250040/in/album-72157624640655996/
33. Carter, J.M. et al. (2016). Estimating National Water Use Associated with Unconventional Oil and Gas Development. U.S. Geological Survey, Fact Sheet 2016-3032, May 2016. https://pubs.usgs.gov/fs/2016/3032/fs20163032.pdf
34. Maupin, M.A. et al. (2014). Estimated Water Use in the United States in 2010. U.S. Geological Survey, Circular 1405. https://pubs.usgs.gov/circ/1405/
35. U.S. Environmental Protection Agency – Water & Energy Efficiency by Sectors: Oil Refineries. https://www3.epa.gov/region9/waterinfrastructure/oilrefineries.html
36. Veil, J. (2015). U.S. Produced Water Volumes and Management Practices in 2012. http://www.veilenvironmental.com/publications/pw/prod_water_volume_2012.pdf
37. "Management of Oil Field Wastes." P. Wright, Oil & Gas Law Report, March 29, 2013. http://www.oilandgaslawreport.com/2013/03/29/management-of-oil-field-wastes/
38. U.S. Pipeline and Hazardous Materials Safety Administration (2012). Leak Detection Study, DTPH56-11-D-000001. https://www.phmsa.dot.gov/sites/phmsa.dot.gov/files/docs/technical-resources/pipeline/16691/leak-detection-study.pdf
39. U.S. Pipeline and Hazardous Materials Safety Administration – PHMSA Pipeline Technical Resources. https://primis.phmsa.dot.gov/ptr.htm
40. Federal Energy Regulatory Commission – What FERC Does. https://www.ferc.gov/about/ferc-does.asp

Part 16: Oil Refining and Gas Processing

1. File:Crude Oil Distillation-en. Wikimedia Commons users Psarianos & Theresa Knott. https://commons.wikimedia.org/wiki/File:Crude_Oil_Distillation-en.svg. Reproduced according to a CC BY-SA 3.0 license: https://creativecommons.org/licenses/by-sa/3.0/deed.en
2. Centre for Industry Education Collaboration, University of York (2014). Cracking and related refinery processes. The Essential Chemical Industry – online. http://www.essentialchemicalindustry.org/processes/cracking-isomerisation-and-reforming.html
3. Kokayeff, P. et al. (2014). Hydrotreating in Petroleum Processing. In: Treese, S., Jones, D., Pujado, P. (eds). Handbook of Petroleum Processing. Springer, Cham. https://link.springer.com/referenceworkentry/10.1007%2F978-3-319-05545-9_4-1
4. U.S. Energy Information Administration (2013). Alkylation is an important source for octane in gasoline. Today in Energy, February 13, 2013. https://www.eia.gov/todayinenergy/detail.php?id=9971
5. U.S. Environmental Protection Agency – Gasoline Standards: Gasoline Reid Vapor Pressure. https://www.epa.gov/gasoline-standards/gasoline-reid-vapor-pressure
6. U.S. Energy Information Administration – Biofuels: Ethanol and Biodiesel Explained – Use of Ethanol. https://www.eia.gov/energyexplained/index.cfm?page=biofuel_ethanol_use
7. U.S. Energy Information Administration – Oil: Crude and Petroleum Products Explained – Refining Crude Oil. https://www.eia.gov/energyexplained/index.cfm?page=oil_refining
8. U.S. Energy Information Administration – Petroleum & Other Liquids: U.S. Product Supplied, Total Crude Oil and Petroleum Products. https://www.eia.gov/dnav/pet/pet_cons_psup_dc_nus_mbblpd_a.htm
9. U.S. Energy Information Administration – U.S. Natural Gas Gross Withdrawals. https://www.eia.gov/dnav/ng/hist/n9010us2a.htm

Petroleum and the Environment References

[10] U.S. Energy Information Administration – Natural Gas Annual Respondent Query System, EIA-757: Natural Gas Processing Capacity by Plant, Data through 2014. https://www.eia.gov/cfapps/ngqs/ngqs.cfm?f_report=RP9

[11] U.S. Energy Information Administration – U.S. Dry Natural Gas Production. https://www.eia.gov/dnav/ng/hist/n9070us2A.htm

[12] North American Energy Standards Board. https://naesb.org/

[13] Penn State College of Earth and Mineral Sciences, e-Education Institute – Petroleum Processing: Natural Gas Composition and Specifications. https://www.e-education.psu.edu/fsc432/content/natural-gas-composition-and-specifications

[14] Rufford, T.E. et al. (2012). The removal of CO_2 and N_2 from natural gas: A review of conventional and emerging process technologies. J. Pet. Sci. Eng., 94-95, 123-154. https://www.sciencedirect.com/science/article/pii/S0920410512001581

[15] Sep-Pro Systems – Nitrogen Rejection Units. http://www.sepprosystems.com/Nitrogen_Rejection_Units.html

[16] U.S. Energy Information Administration (2006). Natural Gas Processing: The Crucial Link between Natural Gas Production and Its Transportation to Market. Accessed at: http://www.dnr.louisiana.gov/assets/docs/oilgas/naturalgas/ngprocess_20060131.pdf

[17] U.S. Energy Information Administration – U.S. Energy Mapping System. https://www.eia.gov/state/maps.php

[18] U.S. Department of Energy (2017). Natural Gas Liquids Primer, with a Focus on the Appalachian Region. https://www.energy.gov/sites/prod/files/2017/12/f46/NGL%20Primer.pdf

[19] U.S. Environmental Protection Agency (2017). Inventory of U.S. Greenhouse Gas Emissions and Sinks: 1990-2015. https://www.epa.gov/ghgemissions/inventory-us-greenhouse-gas-emissions-and-sinks-1990-2015

[20] Global CCS Institute – Projects Database: Large-scale CCS facilities. https://www.globalccsinstitute.com/projects/large-scale-ccs-projects

Part 17: Non-Fuel Products of Oil and Gas

[1] U.S. Energy Information Administration – How much oil is used to make plastic? https://www.eia.gov/tools/faqs/faq.php?id=34&t=6

[2] U.S. Energy Information Administration (2015). Growing U.S. HGL production spurs petrochemical industry investment, Today in Energy, January 29, 2015. https://www.eia.gov/todayinenergy/detail.php?id=19771

[3] U.S. Energy Information Administration – U.S. Refinery Net Production of Naphtha for Petrochemical Feedstock Use. https://www.eia.gov/dnav/pet/hist/LeafHandler.ashx?n=PET&s=MPNRX_NUS_1&f=M

[4] Plotkin, J.S. (2016). Beyond the Ethylene Steam Cracker. American Chemical Society. https://www.acs.org/content/acs/en/pressroom/cutting-edge-chemistry/beyond-the-ethylene-steam-cracker.html

[5] Mitsubishi Chemical Techno-Research, LPG Conference, March 7, 2017. http://www.lpgc.or.jp/corporate/information/program5_Japan2.pdf

[6] Plotkin, J.S. (2015). The Propylene Gap: How Can It Be Filled? American Chemical Society. https://www.acs.org/content/acs/en/pressroom/cutting-edge-chemistry/the-propylene-gap-how-can-it-be-filled.html

[7] U.S. Geological Survey (2018). Nitrogen (Fixed) – Ammonia. Mineral Commodity Summaries, 2018. https://minerals.usgs.gov/minerals/pubs/commodity/nitrogen/mcs-2018-nitro.pdf

[8] Hess, J. et al. (2011). Petroleum and Health Care: Evaluating and Managing Health Care's Vulnerability to Petroleum Supply Shifts. Am. J. Pub. Health, 101(9), 1568-1579. https://ajph.aphapublications.org/doi/10.2105/AJPH.2011.300233

[9] American Cleaning Institute – Soaps & Detergents: Chemistry (Surfactants). https://www.cleaninginstitute.org/clean_living/soaps__detergents_chemistry_2.aspx

[10] U.S. Department of Agriculture, Economic Research Service – Fertilizer Use and Price, 2018. https://www.ers.usda.gov/data-products/fertilizer-use-and-price.aspx

[11] U.S. Energy Information Administration (2017). First world-scale, greenfield nitrogenous fertilizer plant opened in over 25 years, Natural Gas Weekly Update, May 4, 2017. https://www.eia.gov/naturalgas/weekly/archivenew_ngwu/2017/05_04/

[12] Penn State Extension – Gas, Cracker, Polymer, Pellets – Ethane's Journey to Plastics. Shale Gas Webinar, September 21, 2017. https://extension.psu.edu/gas-cracker-polymer-pellets-ethane-s-journey-to-plastics

[13] U.S Energy Information Administration – Natural Gas Gross Withdrawals and Production. https://www.eia.gov/dnav/ng/ng_prod_sum_a_EPG0_FGW_mmcf_a.htm

[14] U.S. Energy Information Administration (2017). Appalachian natural gas processing capacity key to increasing natural gas, NGPL production. Today in Energy, August 29, 2017. https://www.eia.gov/todayinenergy/detail.php?id=32692

[15] U.S. Energy Information Administration (2016). Short-term Outlook for Hydrocarbon Gas Liquids, https://www.eia.gov/outlooks/steo/special/supplements/2016/hgl/pdf/2016_sp_01.pdf

[16] Shell – Pennsylvania Chemicals Project: Frequently Asked Questions. http://www.shell.us/about-us/projects-and-locations/pennsylvania-chemicals-project/appalachian-petrochemical-project-frequently-asked-questions.html

[17] File:Top dressing winter wheat – geograph.org.uk – 1740985. Wikimedia Commons, Author: Michael Trolove. https://commons.wikimedia.org/wiki/File:Top_dressing_winter_wheat_-_geograph.org.uk_-_1740985.jpg Reproduced according to a CC BY-SA 2.0 license: https://creativecommons.org/licenses/by-sa/2.0/deed.en

[18] Photo by Airman Valerie Monroy, U.S. Air Force. https://www.dvidshub.net/image/2485461/dea-host-national-prescription-drug-take-back

[19] Vaseline – What is Petroleum Jelly, and What is it Used For? http://www.vaseline.us/skin-health-care/what-is-petroleum-jelly.html

[20] Connelly, D. (2014). A history of aspirin. Clinical Pharmacist, September 2014, 6(7), online. https://www.pharmaceutical-journal.com/news-and-analysis/infographics/a-history-of-aspirin/20066661.article

[21] U.S. Geological Survey (2018). Sulfur. Mineral Commodity Summaries, 2018. https://minerals.usgs.gov/minerals/pubs/commodity/sulfur/mcs-2017-sulfu.pdf

[22] Apodaca, L.E. (2017). Sulfur. 2015 Minerals Yearbook, U.S. Geological Survey. https://minerals.usgs.gov/minerals/pubs/commodity/sulfur/myb1-2015-sulfu.pdf

[23] U.S. Geological Survey (2018). Helium. Mineral Commodity Summaries, 2018. https://minerals.usgs.gov/minerals/pubs/commodity/helium/mcs-2018-heliu.pdf

[24] File:MRI-Philips.JPG. Wikimedia Commons user Jan Ainali. https://commons.wikimedia.org/wiki/File:MRI-Philips.JPG. Reproduced according to a CC BY 3.0 license: https://creativecommons.org/licenses/by/3.0/deed.en

[25] Barnes, D.K.A. et al. (2009). Accumulation and fragmentation of plastic debris in global environments. Phil. Trans. R. Soc. B, 364, 1985-1998. http://rstb.royalsocietypublishing.org/content/364/1526/1985

Petroleum and the Environment References

26. Ryan, P.G. (1990). The effects of ingested plastic and other marine debris on seabirds. NOAA Technical Memo NOAA-TM-NMFS-SWFSC-154. https://swfsc.noaa.gov/publications/TM/SWFSC/NOAA-TM-NMFS-SWFSC-154_P623.PDF
27. U.S. Environmental Protection Agency – Nutrient Pollution: The Sources and Solutions: Agriculture. https://www.epa.gov/nutrientpollution/sources-and-solutions-agriculture
28. U.S. Environmental Protection Agency – Emergency Response: Sorbents. EPA Archive, January 19, 2017. https://19january2017snapshot.epa.gov/emergency-response/sorbents_.html
29. U.S. Environmental Protection Agency – National Pollutant Discharge Elimination System. https://www.epa.gov/npdes
30. File:Silt fence EPA. Obtained from Wikimedia Commons. Public Domain. https://commons.wikimedia.org/wiki/File:Silt_fence_EPA.jpg
31. Rochelle, G.T. (2009). Amine Scrubbing for CO_2 capture. Science, 325, 1652-1654. http://science.sciencemag.org/content/325/5948/1652
32. "Petra Nova, World's Largest Post-Combustion Carbon Capture Project, Begins Commerical Operation." U.S. Department of Energy, Office of Fossil Energy, January 11, 2017. https://energy.gov/fe/articles/petra-nova-world-s-largest-post-combustion-carbon-capture-project-begins-commercial
33. Container Recycling Institute – Plastic Facts and Statistics. http://www.container-recycling.org/index.php/factsstatistics/plastic
34. Plastic Film Recycling – Learn What's Recyclable. https://www.plasticfilmrecycling.org/recycling-bags-and-wraps/plastic-film-education-individuals/learn-whats-recyclable/
35. American Chemistry Council – What Plastics Can Become. https://www.recycleyourplastics.org/consumers/kids-recycling/plastics-can-become/
36. U.S. Environmental Protection Agency (2016). Advancing Sustainable Materials Management: 2014 Fact Sheet. https://www.epa.gov/sites/production/files/2016-11/documents/2014_smmfactsheet_508.pdf

Part 18: Air Quality Impacts of Oil and Gas

1. World Health Organization – Ambient (outdoor) air quality and health. http://www.who.int/mediacentre/factsheets/fs313/en/
2. U.S. National Park Service – Where Does Air Pollution Come From? https://www.nps.gov/subjects/air/sources.htm
3. Bernice 1 and 2 wells with moisture flare – Evanson Place – Arnegard North Dakota – 2013-07-04. Flickr user Tim Evanson. https://www.flickr.com/photos/23165290@N00/9287130523/. Reproduced according to a CCBY-SA 2.0 license: https://creativecommons.org/licenses/by-sa/2.0/
4. U.S. Environmental Protection Agency – Air Emissions Sources. https://www.epa.gov/air-emissions-inventories/air-emissions-sources
5. U.S. Environmental Protection Agency – What are Hazardous Air Pollutants? https://www.epa.gov/haps/what-are-hazardous-air-pollutants
6. U.S. Environmental Protection Agency – Overview of the Clean Air Act and Air Pollution. https://www.epa.gov/clean-air-act-overview
7. U.S. Environmental Protection Agency – The Clean Air Act in a Nutshell: How it Works, March 2013. https://www.epa.gov/sites/production/files/2015-05/documents/caa_nutshell.pdf
8. U.S. Environmental Protection Agency – New Source Performance Standards. https://www.epa.gov/stationary-sources-air-pollution/new-source-performance-standards
9. U.S. Environmental Protection Agency – Air Emissions Inventories: National Summary of VOC Emissions. https://www3.epa.gov/cgi-bin/broker?polchoice=VOC&_debug=0&_service=data&_program=dataprog.national_1.sas
10. Schade, G.W. and Roest, G. (2016). Analysis of non-methane hydrocarbon data from a monitoring station affected by oil and gas development in the Eagle Ford shale, Texas. Elem. Sci. Anth., 4, 96. https://www.elementascience.org/articles/10.12952/journal.elementa.000096/
11. Texas Commission on Environmental Quality – Air Quality: Eagle Ford Shale Geological Area. https://www.tceq.texas.gov/airquality/eagleford/eagle-ford-main
12. U.S. Environmental Protection Agency – Enforcement: Civil Cases and Settlements. https://cfpub.epa.gov/enforcement/cases/
13. U.S. Environmental Protection Agency – Enforcement: Petroleum Refinery National Case Results. https://www.epa.gov/enforcement/petroleum-refinery-national-case-results
14. U.S. Environmental Protection Agency – Clear Air Act Standards and Guidelines for Petroleum Refineries and Distribution Industry. https://www.epa.gov/stationary-sources-air-pollution/clean-air-act-standards-and-guidelines-petroleum-refineries-and
15. U.S. Environmental Protection Agency (2011). Toxic Release Inventory (TRI) Program: Lifting of Administrative Stay for Hydrogen Sulfide. https://www.epa.gov/toxics-release-inventory-tri-program/lifting-administrative-stay-hydrogen-sulfide
16. U.S. Environmental Protection Agency – AP 42: Compilation of Air Emission Factors, Fifth Edition, Volume 1, Chapter 5: Petroleum Industry. https://www3.epa.gov/ttn/chief/ap42/ch05/index.html
17. U.S. Environmental Protection Agency – Reducing Emissions of Hazardous Air Pollutants. https://www.epa.gov/haps/reducing-emissions-hazardous-air-pollutants
18. "Oil Refiners to Reduce Air Pollution at Six Refineries Under Settlement with EPA and Department of Justice." U.S. Department of Justice Press Release, July 18, 2016. https://www.justice.gov/opa/pr/oil-refiners-reduce-air-pollution-six-refineries-under-settlement-epa-and-department-justice
19. U.S. Environmental Protection Agency (2010). Available and emerging technologies for reducing greenhouse gas emissions from the petroleum refining industry. https://www.epa.gov/sites/production/files/2015-12/documents/refineries.pdf
20. Bay Area Air Quality Management District (2015). Petroleum Refinery Emissions Reduction Strategy: Workshop Report, September 2015. http://www.baaqmd.gov/~/media/files/communications-and-outreach/community-outreach/refinery-rules/workshop_report_final-pdf.pdf
21. Sage Environmental Consulting for the American Fuel and Petrochemical Manufacturers (2015). Historical Air Emissions from United States Petroleum Refineries. https://www.afpm.org/uploadedFiles/Content/documents/Sage-Report.pdf
22. Texas Commission on Environmental Quality (2016). APWL [Air Pollution Watch List] Proposed Change Document: Delisting – Benzene, Galena Park, TX. https://www.tceq.texas.gov/assets/public/implementation/tox/apwlproposal/oct16/1206document.pdf
23. File:Houston Ship Channel Galena. U.S. Army Corps of Engineers. Accessed via Wikimedia Commons. Public Domain. https://commons.wikimedia.org/wiki/File:Houston_Ship_Channel_Galena.jpg

Petroleum and the Environment References

24. U.S. Environmental Protection Agency – Air Quality – National Summary. https://www.epa.gov/air-trends/air-quality-national-summary
25. U.S. Environmental Protection Agency (2017). AirTrends Annual Report: Unhealthy Air Quality Days Trending Down. https://gispub.epa.gov/air/trendsreport/2017/#unhealthy_aq_days
26. "EPA Takes Final Step in Phaseout of Leaded Gasoline." U.S. Environmental Protection Agency Press Release, January 29, 1996. https://archive.epa.gov/epa/aboutepa/epa-takes-final-step-phaseout-leaded-gasoline.html
27. U.S. Federal Aviation Administration – About Aviation Gasoline. https://www.faa.gov/about/initiatives/avgas/
28. U.S. Environmental Protection Agency – Lead Trends. https://www.epa.gov/air-trends/lead-trends
29. U.S. Environmental Protection Agency – Table of Historical Sulfur Dioxide National Ambient Air Quality Standards (NAAQS). https://www.epa.gov/so2-pollution/table-historical-sulfur-dioxide-national-ambient-air-quality-standards-naaqs
30. U.S. Environmental Protection Agency – Acid Rain Program: Overview. https://www.epa.gov/airmarkets/acid-rain-program
31. U.S. Environmental Protection Agency (2017). AirTrends Annual Report: Air Quality Improves as America Grows. https://gispub.epa.gov/air/trendsreport/2017/#highlights
32. U.S. Environmental Protection Agency – Table of Historical Particulate Matter (PM) National Ambient Air Quality Standards. https://www.epa.gov/pm-pollution/table-historical-particulate-matter-pm-national-ambient-air-quality-standards-naaqs
33. U.S. Environmental Protection Agency – Table of Historical Ozone National Ambient Air Quality Standards. https://www.epa.gov/ozone-pollution/table-historical-ozone-national-ambient-air-quality-standards-naaqs
34. U.S. Environmental Protection Agency – Air Pollutant Emissions Trends Data. https://www.epa.gov/air-emissions-inventories/air-pollutant-emissions-trends-data
35. U.S. Environmental Protection Agency – Regulations to Reduce Mobile Source Pollution. https://www.epa.gov/mobile-source-pollution/regulations-reduce-mobile-source-pollution
36. U.S. Environmental Protection Agency – Ambient Concentrations of Benzene. https://cfpub.epa.gov/roe/documents/BenzeneConcentrations.pdf
37. HDR: Los Angeles Skyline. Flickr user Al Pavangkanan. https://www.flickr.com/photos/drtran/2186120627. Reproduced according to a CC BY 2.0 license: https://creativecommons.org/licenses/by/2.0/
38. U.S. Environmental Protection Agency – Air Pollution: Current and Future Challenges. https://www.epa.gov/clean-air-act-overview/air-pollution-current-and-future-challenges
39. U.S. Environmental Protection Agency – Benefits and Costs of the Clean Air Act 1990-2020, the Second Prospective Study. https://www.epa.gov/clean-air-act-overview/benefits-and-costs-clean-air-act-1990-2020-second-prospective-study

Part 19: Methane Emissions in the Oil and Gas Industry

1. U.S. Environmental Protection Agency – Greenhouse Gas Emissions: Overview of Greenhouse Gases. https://www.epa.gov/ghgemissions/overview-greenhouse-gases
2. U.S. Energy Information Administration – Electric Power Monthly, Table 1.1 – Net Generation by Energy Source: Total (All Sectors), 2007-December 2017. https://www.eia.gov/electricity/monthly/epm_table_grapher.php?t=epmt_1_1
3. National Energy Technology Laboratory (2013). Cost and Performance Baseline for Fossil Energy Plants, Volume 1: Bituminous Coal and Natural Gas to Electricity, Revision 2a, September 2013. https://www.netl.doe.gov/File%20Library/Research/Energy%20Analysis/OE/BitBase_FinRep_Rev2a-3_20130919_1.pdf
4. U.S. Environmental Protection Agency (2017). Inventory of U.S. Greenhouse Gas Emissions and Sinks: 1990-2015. https://www.epa.gov/ghgemissions/inventory-us-greenhouse-gas-emissions-and-sinks-1990-2015
5. Schmidt, G. (2004). Methane: A Scientific Journey from Obscurity to Super-Stardom. NASA Research Features. https://www.giss.nasa.gov/research/features/200409_methane/
6. United Nations Framework Convention on Climate Change – National Reports. https://unfccc.int/topics/mitigation/workstreams/nationally-appropriate-mitigation-actions/national-reports
7. U.S. Environmental Protection Agency – Greenhouse Gas Reporting Program (GHGRP). https://www.epa.gov/ghgreporting
8. U.S. Environmental Protection Agency (2013). Petroleum and Natural Gas Systems: 2011 Data Summary. https://www.epa.gov/sites/production/files/2016-02/documents/supporting-info-2011-data-summary.pdf
9. Heath, G. et al. (2015). Estimating U.S. Methane Emissions from the Natural Gas Supply Chain: Approaches, Uncertainties, Current Estimates, and Future Studies. Joint Institute for Strategic Energy Analysis, Technical Report NREL/TP-6A50-62820. https://www.nrel.gov/docs/fy16osti/62820.pdf
10. Zavala-Araiza, D. et al. (2015). Toward a Functional Definition of Methane Super-Emitters: Application to Natural Gas Production Sites. Environ. Sci. Technol., 49(13), 8167-8174. http://pubs.acs.org/doi/full/10.1021/acs.est.5b00133
11. Lattanzio, R.K. (2018). Methane and Other Air Pollution Issues in Natural Gas Systems. Congressional Research Service Report R42986
12. "EDF Announces Satellite Mission to Locate and Measure Methane Emissions." Environmental Defense Fund Press Release, April 11, 2018. https://www.edf.org/media/edf-announces-satellite-mission-locate-and-measure-methane-emissions
13. U.S. Environmental Protection Agency – U.S. Greenhouse Gas Inventory Report Archive. https://www.epa.gov/ghgemissions/us-greenhouse-gas-inventory-report-archive
14. Environmental Defense Fund (2017) – Methane Research: The 16 Study Series. https://www.edf.org/sites/default/files/methane_studies_fact_sheet.pdf
15. Lyon, D. et al. (2016). Aerial Surveys of Elevated Hydrocarbon Emissions from Oil and Gas Production Sites. Environ. Sci. Technol., 50(9), 4877-4886. https://pubs.acs.org/doi/abs/10.1021/acs.est.6b00705
16. Townsend-Small, A. et al. (2015). Integrating Source Apportionment Tracers into a Bottom-up Inventory of Methane Emissions in the Barnett Shale Hydraulic Fracturing Region. Environ. Sci. Technol., 49(13), 8175-8182. https://pubs.acs.org/doi/abs/10.1021/acs.est.5b00057

Petroleum and the Environment References

[17] Zavala-Araiza, D. et al. (2015). Reconciling divergent estimates of oil and gas methane emissions. Proc. Natl. Acad. Sci., 112(51), 15597-15602. http://www.pnas.org/content/112/51/15597.full

Part 20: Mitigating and Regulating Methane Emissions

[1] U.S. Environmental Protection Agency – Natural Gas STAR Program: Overview of the Oil and Natural Gas Industry. https://www.epa.gov/natural-gas-star-program/overview-oil-and-natural-gas-industry

[2] File:Flaring extra Gas in the Bakken.JPG. Wikimedia Commons user Joshua Doubek. https://commons.wikimedia.org/wiki/File:Flaring_extra_Gas_in_the_Bakken.JPG Reproduced according to a CC BY-SA 3.0 license: https://creativecommons.org/licenses/by-sa/3.0/deed.en

[3] U.S. Environmental Protection Agency – Understanding Global Warming Potentials. https://www.epa.gov/ghgemissions/understanding-global-warming-potentials

[4] U.S. Environmental Protection Agency – Ozone Pollution. https://www.epa.gov/ozone-pollution

[5] American Petroleum Institute (2017). Natural Gas, Oil Industry Launch Environmental Partnership to Accelerate Reductions in Methane, VOCs. December, 2017. http://www.api.org/news-policy-and-issues/news/2017/12/04/natural-gas-oil-environmental-partnership-accelerate-reductions-methane-vocs

[6] U.S. Environmental Protection Agency – Controlling Air Pollution from the Oil and Natural Gas Industry: New Source Performance Standards and Permitting Requirements. https://www.epa.gov/controlling-air-pollution-oil-and-natural-gas-industry/new-source-performance-standards-and

[7] National Conference of State Legislatures (2014). State Methane Policies. http://www.ncsl.org/research/environment-and-natural-resources/state-methane-policies.aspx

[8] International Petroleum Industry Environmental Conservation Association (2014). Green Completions. http://www.ipieca.org/resources/energy-efficiency-solutions/units-and-plants-practices/green-completions/

[9] U.S. Environmental Protection Agency (2012). Overview of Final Amendments to Air Regulations for the Oil and Gas Industry. https://www.epa.gov/sites/production/files/2016-09/documents/natural_gas_transmission_fact_sheet_2012.pdf

[10] U.S. Environmental Protection Agency – Actions and Notices about Oil and Natural Gas Air Pollution Standards. https://www.epa.gov/controlling-air-pollution-oil-and-natural-gas-industry/actions-and-notices-about-oil-and-natural-gas

[11] U.S. Energy Information Administration (2016). Natural gas flaring in North Dakota has declined sharply since 2014. Today in Energy, June 13, 2016. https://www.eia.gov/todayinenergy/detail.php?id=26632

[12] Roy Luck – Crosstex gas processing facility. Taken on May 9, 2009. Flickr. https://www.flickr.com/photos/royluck/3519496307. Reproduced according to a CC BY 2.0 license. https://creativecommons.org/licenses/by/2.0/

[13] U.S. Bureau of Land Management (2016). Waste Prevention, Production Subject to Royalties, and Resource Conservation. Federal Register, 81, p. 83008-83089, November 18, 2016. https://www.federalregister.gov/documents/2016/11/18/2016-27637/waste-prevention-production-subject-to-royalties-and-resource-conservation

[14] "EPA Releases First-Ever Standards to Cut Methane Emissions from the Oil and Gas Sector." U.S. Environmental Protection Agency press release, May 12, 2016. https://archive.epa.gov/epa/newsreleases/epa-releases-first-ever-standards-cut-methane-emissions-oil-and-gas-sector.html

[15] Clark Hill PLC (2017). Court Vacates EPA Stay of Methane Emissions Final Rule. Lexology, September 29, 2017. https://www.lexology.com/library/detail.aspx?g=009a6130-6ef8-49f4-af57-e4e2c925fa33

[16] California Air Resources Board (2017). CARB approves rule for monitoring and repairing methane leaks from oil and gas facilities. News Release #17-18, March 23, 2017. https://www.arb.ca.gov/newsrel/newsrelease.php?id=907

[17] Pennsylvania Department of Environmental Protection – How Pennsylvania is Regulating Methane from the Oil and Gas Industry. http://files.dep.state.pa.us/Air/AirQuality/AQPortalFiles/Permits/gp/MethaneRegulations.pdf

[18] U.S. Environmental Protection Agency – EPA's Voluntary Methane Programs for the Oil and Natural Gas Industry. https://www.epa.gov/natural-gas-star-program

[19] U.S. Environmental Protection Agency – AgSTAR: Biogas Recovery in the Agricultural Sector. https://www.epa.gov/agstar

Part 21: Regulation of Oil and Gas Operations

[1] Railroad Commission of Texas - Railroad Commission Milestones. http://www.rrc.state.tx.us/about-us/history/milestones/

[2] Interstate Oil & Gas Compact Commission – Our History. http://iogcc.publishpath.com/history

[3] U.S. Environmental Protection Agency – Regulatory Information by Sector: Oil and Gas Extraction Sector. https://www.epa.gov/regulatory-information-sector/oil-and-gas-extraction-sector-naics-211

[4] U.S. Occupational Safety and Health Administration – OSHA Law & Regulations. https://www.osha.gov/law-regs.html

[5] Interstate Oil & Gas Compact Commission – Member States. http://iogcc.ok.gov/member-states

[6] Mapchart.net. https://mapchart.net/usa.html

[7] Interstate Oil & Gas Compact Commission – Interstate Oil and Gas Commission Charter. http://iogcc.publishpath.com/charter

[8] Interstate Oil & Gas Compact Commission – Summary of State Statutes and Regulations. http://iogcc.publishpath.com/state-statutes

[9] Ground Water Protection Council (2009). State Oil and Natural Gas Regulations Designed to Protect Water Resources. Prepared for the U.S. Department of Energy's National Energy Technology Laboratory, May 2009. http://www.gwpc.org/sites/default/files/state_oil_and_gas_regulations_designed_to_protect_water_resources_0.pdf

[10] U.S. Office of Natural Resources Revenue – Production Data. https://www.onrr.gov/About/production-data.htm

[11] U.S. Energy Information Administration – Crude Oil Production. https://www.eia.gov/dnav/pet/pet_crd_crpdn_adc_mbblpd_a.htm

Petroleum and the Environment References

12. U.S. Energy Information Administration – U.S. Dry Natural Gas Production. https://www.eia.gov/dnav/ng/hist/n9070us2m.htm
13. U.S. Geological Survey – The National Map: Federal Lands and Indian Reservations. https://nationalmap.gov/small_scale/printable/fedlands.html
14. U.S. Environmental Protection Agency – Regulatory Information by Sector: Oil and Gas Extraction Sector. https://www.epa.gov/regulatory-information-sector/oil-and-gas-extraction-sector-naics-211
15. U.S. Environmental Protection Agency – Reduced Emission Completions for Hydraulically Fractured Natural Gas Wells. https://www.epa.gov/natural-gas-star-program/reduced-emission-completions-hydraulically-fractured-natural-gas-wells
16. U.S. Environmental Protection Agency – Actions and Notices about Oil and Natural Gas Air Pollution Standards. https://www.epa.gov/controlling-air-pollution-oil-and-natural-gas-industry/actions-and-notices-about-oil-and-natural-gas
17. "Model Advances Analysis of Offshore BOP Closure during Extreme Pressure and Flow Conditions." U.S. Bureau of Safety and Environmental Enforcement press release, February 17, 2017. Accessed at: https://www.flickr.com/photos/bseegov/33036282232
18. U.S. Environmental Protection Agency (2015). Minimizing and Managing Potential Impacts of Injection-Induced Seismicity from Class II Disposal Wells: Practical Approaches. https://www.epa.gov/sites/production/files/2015-08/documents/induced-seismicity-201502.pdf
19. Bureau of Land Management – Oil and Gas: Regulations, Onshore Orders and Notices to Lessees. https://www.blm.gov/programs/energy-and-minerals/oil-and-gas/operations-and-production/onshore-orders
20. "BLM Rescinds Rule on Hydraulic Fracturing." U.S. Bureau of Land Management Press Release, December 28, 2017. https://www.blm.gov/press-release/blm-rescinds-rule-hydraulic-fracturing
21. National Park Service (2015). Cost-Benefit and Regulatory Flexibility Analyses: U.S. Department of the Interior, National Park Service, Proposed Revisions to 36 CFR Part 9, Subpart B. https://www.nps.gov/subjects/energyminerals/upload/2015-09-21-NPS-Non-Federal-Oil-and-Gas-CBA-FINAL.pdf
22. Code of Federal Regulations – Title 36, Chapter I, Part 6, Subpart B: Non-Federal Oil and Gas Rights.
23. U.S. Environmental Protection Agency – National Environmental Policy Act. https://www.epa.gov/nepa
24. U.S. Bureau of Ocean Energy Management – Outer Continental Shelf. https://www.boem.gov/Outer-Continental-Shelf/
25. Cameron Jr., B. and Matthews, T. (2016). OCS Regulatory Framework. U.S. Bureau of Ocean Energy Management report BOEM 2016-014. https://www.boem.gov/OCS-Regulatory-Framework/
26. National Academy of Engineering and National Research Council (2012). Macondo Well Deepwater Horizon Blowout: Lessons for Improving Offshore Drilling Safety. Washington, DC: The National Academies Press. https://www.nap.edu/catalog/13273/macondo-well-deepwater-horizon-blowout-lessons-for-improving-offshore-drilling
27. U.S. Bureau of Safety and Environmental Enforcement – Regulations, https://www.bsee.gov/guidance-and-regulations/regulations
28. U.S. Pipeline and Hazardous Materials Safety Administration – State Programs Overview. https://www.phmsa.dot.gov/working-phmsa/state-programs/state-programs-overview
29. Pless, J. (2011). Making State Gas Pipelines Safe and Reliable: An Assessment of State Policy. National Conference of State Legislatures, March 2011. http://www.ncsl.org/research/energy/state-gas-pipelines-federal-and-state-responsibili.aspx
30. U.S. Pipeline and Hazardous Materials Safety Administration – Pipeline Technical Resources. https://primis.phmsa.dot.gov/ptr.htm
31. U.S. Pipeline and Hazardous Materials Safety Administration – Underground Natural Gas Storage. https://primis.phmsa.dot.gov/ung/index.htm
32. U.S. Environmental Protection Agency – Clean Air Act Standards and Guidelines for Petroleum Refineries and Distribution Industry. https://www.epa.gov/stationary-sources-air-pollution/clean-air-act-standards-and-guidelines-petroleum-refineries-and
33. Federal Energy Regulatory Commission – What FERC does. https://www.ferc.gov/about/ferc-does.asp
34. Shea, D. et al. (2015). Transporting Crude Oil by Rail: State and Federal Action. National Conference of State Legislatures, October 30, 2015. http://www.ncsl.org/research/energy/transporting-crude-oil-by-rail-state-and-federal-action.aspx
35. Durkay, J. (2017). State Renewable Portfolio Standards and Goals. National Conference of State Legislatures, August 1, 2017. http://www.ncsl.org/research/energy/renewable-portfolio-standards.aspx
36. U.S. Environmental Protection Agency – Renewable Fuel Standard Program. https://www.epa.gov/renewable-fuel-standard-program
37. U.S. Environmental Protection Agency – Final Renewable Fuel Standards for 2018, and the Biomass-based Diesel Volume for 2019. https://www.epa.gov/renewable-fuel-standard-program/final-renewable-fuel-standards-2018-and-biomass-based-diesel-volume
38. U.S. Energy Information Administration (2016). Almost all U.S. Gasoline is blended with 10% ethanol. Today in Energy, May 4, 2016. https://www.eia.gov/todayinenergy/detail.php?id=26092
39. U.S. Fish & Wildlife Service – Endangered Species Act: Overview. https://www.fws.gov/endangered/laws-policies/
40. Joy, M.P. and Dimitroff, S.D. (2016). Oil and gas regulation in the United States: overview. Westlaw, June 1, 2016. https://content.next.westlaw.com/Document/I466099551c9011e38578f7ccc38dcbee/View/FullText.html

Part 22: Health and Safety in Oil and Gas Extraction

1. U.S. Bureau of Labor Statistics – Quarterly Census of Employment and Wages. https://www.bls.gov/cew/ (2017 NAICS codes: 211120, 211130, 213111, and 213112)
2. U.S. Bureau of Labor Statistics – Census of Fatal Occupational Injuries. https://www.bls.gov/iif/oshcfoi1.htm
3. Mason, K.L. et al. (2015). Occupational Fatalities During the Oil and Gas Boom – United States, 2003-2013. Centers for Disease Control and Prevention, Morbidity and Mortality Weekly Report, 64(2), 551-554. https://www.cdc.gov/Mmwr/preview/mmwrhtml/mm6420a4.htm
4. National Institute for Occupational Safety and Health (2010). NIOSH Field Effort to Assess Chemical Exposure Risks to Gas and Oil Workers. Fact Sheet, DHHS (NIOSH) Publication No. 2010-130. https://www.cdc.gov/niosh/docs/2010-130/pdfs/2010-130.pdf
5. Kiefer, M. (2013). NIOSH Safety and Health Research in Oil and Gas Extraction. Board of Scientific Counselors Meeting, September 18, 2013. https://www.cdc.gov/niosh/docket/archive/pdfs/niosh-278/nioshbschydraulicfracturing_09182013.pdf

Petroleum and the Environment References

[6] Esswein et al. (2013). Occupational Exposures to Respirable Crystalline Silica During Hydraulic Fracturing. J. Occup. Environ. Hyg., 10(7), 347-356. https://www.ncbi.nlm.nih.gov/pubmed/23679563

[7] NIOSH Research Rounds – Investigators Design Experimental Engineering Control for Silica Dust. https://www.cdc.gov/niosh/research-rounds/resroundsv2n1.html

[8] National Institute for Occupational Safety and Health – Silica. https://www.cdc.gov/niosh/topics/silica/default.html

[9] Occupational Safety and Health Administration (2012). OSHA-NIOSH Hazard Alert: Worker Exposure to Silica During Hydraulic Fracturing. https://www.osha.gov/dts/hazardalerts/hydraulic_frac_hazard_alert.pdf

[10] Occupational Safety and Health Administration – Frequently Asked Questions: Respirable Crystalline Silica Rule. https://www.osha.gov/silica/Silica_FAQs_2016-3-22.pdf

[11] National Institute for Occupational Safety and Health (2016). NIOSH-OSHA Hazard Alert: Health and Safety Risks for Workers Involved in Manual Tank Gauging and Sampling at Oil and Gas Extraction Sites. https://www.cdc.gov/niosh/docs/2016-108/pdfs/2016-108.pdf

[12] California Department of Public Health and National Institute for Occupational Safety and Health (2017). Protecting Oil and Gas Workers from Hydrocarbon Gases and Vapors. Video Pub. No. 2017-158D. https://www.cdc.gov/niosh/docs/video/2017-158d/default.html

[13] Occupational Safety and Health Administration – Occupational Noise Exposure. https://www.osha.gov/SLTC/noisehearingconservation/

[14] Occupational Safety and Health Administration – Occupational Safety and Health Standards: Air Contaminants. Standard Number 1910.1000. https://www.osha.gov/pls/oshaweb/owadisp.show_document?p_table=STANDARDS&p_id=9991

[15] International Agency for Research on Cancer (2012). Chemical Agents and Related Occupations. IARC Monographs on the Evaluation of Carcinogenic Risks to Humans, Volume 100F. http://monographs.iarc.fr/ENG/Monographs/vol100F/

[16] Occupational Safety and Health Administration - Hazard Communication Standard: Safety Data Sheets. https://www.osha.gov/Publications/OSHA3514.html

[17] NORA Oil and Gas Extraction Council, National Institute for Occupational Safety and Health. https://www.cdc.gov/niosh/nora/councils/oilgas/default.html

[18] American Petroleum Institute – Occupational Health and Safety in the Industry. http://www.api.org/oil-and-natural-gas/health-and-safety/health-and-safety-in-the-industry

[19] Association of Energy Service Companies - Industry Resources. http://www.aesc.net/AESC/Industry_Resources/AESC/Industry_Resources/Industry_Resources.aspx?hkey=67c9ecae-7ac2-420f-9ae7-7c3389b27a30

[20] International Association of Drilling Contractors – Health, Safety, and Environmental Case Guidelines. http://www.iadc.org/iadc-hse-case-guidelines/

[21] National STEPS Network. https://www.nationalstepsnetwork.com/

Part 23: Subsurface Data in the Oil and Gas Industry

[1] U.S. Energy Information Administration (2016). Trends in U.S. Oil and Natural Gas Upstream Costs. https://www.eia.gov/analysis/studies/drilling/pdf/upstream.pdf

[2] Bureau of Ocean Energy Management – Record of Decision, Atlantic OCS Region Geological and Geophysical Activities. https://www.boem.gov/Record-of-Decision/

[3] Varhaug, M. (2016). Basic Well Log Interpretation. The Defining Series, Oilfield Review. https://www.slb.com/-/media/Files/resources/oilfield_review/defining_series/Defining-Log-Interpretation.pdf?la=en&hash=3DD25483EEE6EBAA69320AF7612AB9E9C4066993

[4] Schlumberger – 1920s: The First Well Log. https://www.slb.com/about/history/1920s.aspx

[5] AAPGWiki – Overview of Routine Core Analysis. http://wiki.aapg.org/Overview_of_routine_core_analysis

[6] Zihlman, F.N. et al. (2000). Selected Data from Fourteen Wildcat Wells in the National Petroleum Reserve in Alaska. USGS Open File Report 00-200. Core from the well "East Simpson 2", Image no. 0462077. https://certmapper.cr.usgs.gov/data/pubarchives/of00-200/wells/ESIMP2/CORE/ES2CORE.HTM

[7] Society of Petroleum Engineers PetroWiki – Petrophysics. http://petrowiki.org/Petrophysics

[8] "Big Data Growth Continues in Seismic Surveys." K. Boman, Rigzone, September 2, 2015. http://www.rigzone.com/news/oil_gas/a/140418/big_data_growth_continues_in_seismic_surveys

[9] U.S. Geological Survey Core Research Center – Frequently Asked Questions. https://geology.cr.usgs.gov/crc/FAQ.html

[10] Kansas Geological Society & Library – Oil and Gas Well Data. http://www.kgslibrary.com/

[11] Akintomide, A.O. and Dawers, N.H. (2016). Structure of the Northern Margin of the Terrebonne Trough, Southeastern Louisiana: Implications for Salt Withdrawal and Miocene to Holocene Fault Activity. Geological Society of America Abstracts with Programs, 48(7), Paper No. 244-2. https://gsa.confex.com/gsa/2016AM/webprogram/Paper286148.html

[12] Shaw, J. and Shearer, P. (1999). An Elusive Blind-Thrust Fault Beneath Metropolitan Los Angeles. Science, 283, 1516-1518. http://science.sciencemag.org/content/283/5407/1516